Cognitive Radio Networks

Kaigui Bian · Jung-Min Park
Bo Gao

Cognitive Radio Networks

Medium Access Control for Coexistence of Wireless Systems

 Springer

Kaigui Bian
Peking University
Beijing
China

Jung-Min Park
Bo Gao
ECE
Virginia Tech
Blacksburg, VA
USA

ISBN 978-3-319-34355-6   ISBN 978-3-319-07329-3   (eBook)
DOI 10.1007/978-3-319-07329-3

Springer Cham Heidelberg New York Dordrecht London

Preface

During the last few decades, our use of or reliance on the radio spectrum has grown tremendously. Today, we rely on wireless devices and systems to not only enable on-demand, pervasive communications for a large proportion of the population, but also other critical application areas such as scientific and medical research, industrial control and automation, and public safety. As wireless systems and applications continue to proliferate, the demand for precious spectrum resources will continue to grow. For instance, it has been documented that since the release of the latest generation of smart phones, data traffic on some of the cellular networks has increased by over 6,000 %. In the foreseeable future, we expect that the demand for spectrum will continue to increase as new wireless technologies and applications with high data throughput requirements continue to emerge. This insatiable appetite for additional spectrum resources cannot be met by simply allocating new spectrum. The usable capacity of spectrum must be expanded with innovative technologies, regulatory reforms, and removal of market barriers.

Cognitive radio is one of the innovative technologies that has the potential to effectively address the spectrum shortage problem and radically change the way we utilize spectrum. Because of its potential impact, various stakeholders—including regulatory policymakers, wireless device manufacturers, telecommunication operators, and academic researchers—have shown strong interest in it, especially with respect to research and development. Although numerous journal and conference publications, tutorials, and books on cognitive radio have been published in the last few years, the vast majority of them focus on the various physical-layer attributes of the technology. More importantly, these technical publications discuss the cognitive radio in isolation, essentially as a standalone system or network, with little regard for how it may interact with legacy wireless systems or how heterogeneous cognitive radio systems may collaborate with each other. Although this book's main theme is cognitive radio, its specific focus areas are quite different from the existing literature. The primary aim of this book is to provide a comprehensive discussion on how cognitive radio technologies can be employed to enable efficient and harmonious coexistence of homogeneous as well as heterogeneous wireless systems and networks. Because the discussions in the book focus on the problem of coexistence of wireless systems, most of the book's contents relate to the medium access control layer, rather than the physical layer. In other words, the discussions in this book revolve around how cognitive radio technologies can be used to enable various wireless networks to coexist and efficiently share spectrum.

The intended readership of this book includes wireless communications industry researchers and practitioners as well as researchers in academia. The readership is assumed to have background knowledge in wireless communications and networking, although they may have no in-depth knowledge of cognitive radio technologies. The intention of this book is to introduce communication generalists to the technical challenges of the various coexistence techniques and mechanisms as well as solution approaches which are enabled by cognitive radios.

Below, we provide a brief summary of the contents of each chapter.

- Chapter 1: Introduction. This chapter includes an introduction to a few challenging problems related to the medium access control (MAC) layer protocol design for coexistence of cognitive radio networks.
- Chapter 2: Taxonomy of Coexistence Mechanisms. In this chapter, we discuss the background knowledge on existing coexistence mechanisms in wireless networks, and present a taxonomy that classifies the state-of-art research according to various criteria.
- Chapter 3: Rendezvous of Cognitive Radios. This chapter presents two channel hopping-based rendezvous protocols, for clock synchronous and asynchronous cognitive radio networks, respectively.
- Chapter 4: Coexistence-aware Spectrum Sharing for Homogeneous Cognitive Radio Networks. In this chapter, we present an inter base station (BS) Coexistence-Aware Spectrum Sharing protocol for improving the coexistence of infrastructure-based homogeneous cognitive radio networks.
- Chapter 5: Frequency Reuse Over a Single TV White Space Channel. In this chapter, we study the channel sharing problem where multiple network cells are forced to reuse a single TV white-space channel.
- Chapter 6: Channel Assignment for Multi-hop Cognitive Radio Networks. In this chapter, we propose the segment-based channel assignment algorithm for single radio interface, multi-hop cognitive radio networks.
- Chapter 7: Ecology-inspired Coexistence of Heterogeneous Cognitive Radio Networks. This chapter discusses challenges in heterogeneous coexistence mechanisms and proposes a mediation-based spectrum sharing mechanism for coexistence of heterogeneous wireless systems operating over the white-space.

We would like to acknowledge and thank a number of colleagues who have made this book possible. In particular, we would like to acknowledge Dr. Ruiliang Chen at Microsoft and Prof. Xiaoming Li at Peking University. Through collaborative research or discussions, the colleagues mentioned above have provided invaluable inputs that helped shape the contents of this book. We would also like to thank our publishers at Springer, in particular Jessica Lauffer and Charles Glaser, for their helpful guidance and encouragement during the creation of this book.

Contents

About the Authors

Kaigui Bian received his B.S. degree in Computer Science from Peking University in 2001, and received his Ph.D. degree in Computer Engineering from Virginia Tech in 2011. He is currently an Associate Professor in the Institute of Network Computing and Information Systems, School of EECS at Peking University. His research interests include cognitive radio networks, mobile computing, network security, and privacy. He was a visiting scholar at Microsoft Research Asia in 2013. He is a member of the IEEE, the ACM, and the CCF.

Jung-Min Park received a Ph.D. degree in Electrical and Computer Engineering from Purdue University in 2003. He is currently an Associate Professor in the Department of Electrical and Computer Engineering at Virginia Tech, and the site director of a National Science Foundation (NSF) Industry-University Cooperative Research Center (I-UCRC) called Broadband Wireless Access & Applications Center (BWAC). As the site director of BWAC at Virginia Tech, Park is leading several sponsored research projects on wireless networks and network security. He is widely recognized for his pioneering work on enforcement and security problems in cognitive radio networks. His research interests include cognitive radio networks, spectrum sharing technologies, network security and privacy, and applied cryptography. Current or recent research sponsors include the NSF, National Institutes of Health (NIH), Defense Advanced Research Projects Agency (DARPA), Office of Naval Research (ONR), SANS (SysAdmin, Audit, Network Security) Institute, Motorola Solutions, Samsung Electronics, and SCA Techniques. More details on his research interests can be found at http://www.arias.ece. vt.edu and http://www.bwac.wireless.vt.edu/index.html. He is a recipient of the 2014 Virginia Tech College of Engineering Faculty Fellow Award, a 2008 NSF Faculty Early Career Development (CAREER) Award, a 2008 Hoeber Excellence in Research Award, a 1998 AT&T Leadership Award, and a coauthor of a paper that won the Best Paper Award at the 2014 IEEE Symposium on New Frontiers in Dynamic Spectrum Access Networks (DySPAN). He is a senior member of the IEEE and the ACM, and a member of the Korean–American Scientists and Engineers Association (KSEA).

Bo Gao is currently a Ph.D. student in the Department of Electrical and Computer Engineering at Virginia Tech. He received his Bachelor's degree in Electrical Engineering from Beijing Jiaotong University, China in 2006, and his Master's degree in Electrical Engineering from Shanghai Jiaotong University, China in 2009. His research interests include wireless networking, dynamic spectrum access, and network coexistence.

Chapter 1
Introduction

The proliferation of wireless applications operating in unlicensed spectrum bands has resulted in the overcrowding of those bands. In contrast, recent studies have shown that most licensed bands are under-utilized [27]. To address the spectrum shortage problem, the Federal Communications Commission (FCC) is considering the adoption of a new regulatory spectrum management paradigm in which licensed bands are opened up to unlicensed operations [28]. This new regulatory model is often called "opportunistic spectrum sharing (OSS)". In the OSS paradigm, unlicensed users (a.k.a. secondary users) opportunistically access licensed spectrum on a non-interference basis to licensed users (a.k.a. incumbent or primary users).

The cognitive radio (CR) is seen as the key enabling technology for realizing OSS. A cognitive radio has the capability to sense its environment and adapt its mode of operation to achieve its performance objectives [1, 42]. In CR networks, spectrum opportunities, e.g., "TV white spaces" (TVWSs), in licensed bands are identified by the process of spectrum sensing [18]. Once spectrum opportunities are identified, those spectrum resources are mapped into logical channels, where each logical channel is the unit of channel assignment. A CR network composed of secondary users (nodes[1]) equipped with CRs can coexist with incumbent users in licensed bands under the OSS paradigm.

1.1 Coexistence Problems in Wireless Networks

The coexistence of wireless networks can be broadly classified into two categories: *vertical coexistence* and *horizontal coexistence*.

- Vertical coexistence refers to the coexistence of two or more networks that have different priorities to access spectrum. For instance, in CR networks, incumbent users have priority over secondary users when accessing the licensed spectrum bands, which is also called *incumbent coexistence*.

---

[1] We use "node" and "secondary user" interchangeably.

© Springer International Publishing Switzerland 2014
K. Bian et al., *Cognitive Radio Networks*,
DOI: 10.1007/978-3-319-07329-3_1

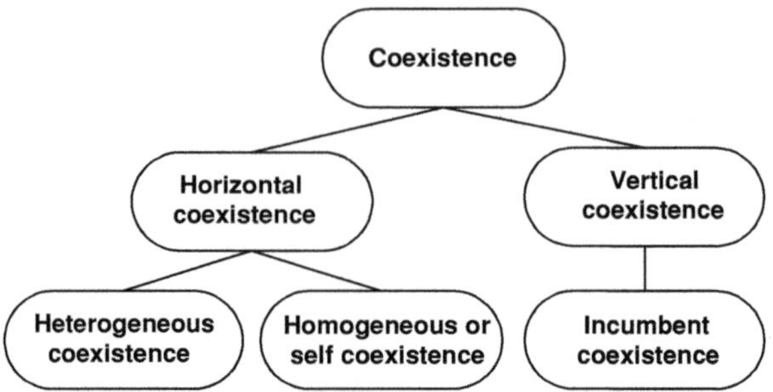

Fig. 1.1 Taxonomy of coexistence in wireless networks

- Horizontal coexistence refers to the coexistence of two or more unlicensed networks that have equal priority to access spectrum.

In the OSS paradigm, the coexistence between incumbent users and secondary users is referred to as *incumbent coexistence*. There exists a significant body of work on incumbent coexistence [14, 16, 17, 32, 89, 111], and it has been attracting significant interest from academia and industry. In contrast, horizontal coexistence has garnered less attention thus far. Horizontal coexistence can be further categorized into:

- *Heterogeneous* coexistence that refers to the coexistence of networks that employ different wireless technologies (e.g., the coexistence between WiFi and Bluetooth [46, 123], the coexistence of heterogeneous wireless networks over TV white space [49]);
- *Homogeneous* coexistence (a.k.a. self coexistence) that refers to the coexistence of networks that employ the same wireless technology (e.g., neighboring CR networks of the same type [12, 64], or neighboring 802.11 hotspots [72]).

Fig. 1.1 illustrates the taxonomy of coexistence problems in wireless networks.

1.2 Medium Access Control Problems for Coexistence of CR Networks

In this book, we investigate a few challenging problems that are related to the medium access control (MAC) layer protocol design for vertical and/or horizontal coexistence in CR networks. Existing MAC layer protocols in conventional wireless networks fail to adequately address the key issues concerning these coexistence problems that emerge in CR networks.

1.2.1 Rendezvous (Control Channel) Establishment

Existing rendezvous protocols can be classified into two approaches: common control channel-based rendezvous [19, 54, 78, 92, 120] and channel hopping (CH)-based rendezvous (a.k.a. parallel rendezvous) [6, 93].

- The use of a *single* common control channel simplifies the rendezvous process but may create a network performance bottleneck and it acts as a single point of failure [6, 74].
- Channel hopping-based rendezvous avoids the drawbacks of the common control channel and creates parallel rendezvous—multiple pairs of nodes are able to rendezvous on different channels simultaneously.

To ensure successful rendezvous (of secondary node pairs that need to establish a control channel) with a high probability, the CH-based rendezvous process needs to support rendezvous in *multiple* channels so that the risk of rendezvous failure due to the appearance of primary user signals is minimized. Ideally, the rendezvous channels should be spread out over all available channels to maximize the chance of rendezvous success.

Most existing CH protocols (e.g., [6, 93]) assumes the clock synchronization, which implies that the communicating nodes are able to align the boundaries of their CH sequences. Meanwhile, it is worth noting that there are CH schemes that enable rendezvous without requiring clock synchronization—e.g., the sequence-based rendezvous protocol [23]. However, this protocol only supports a very limited number of pairwise rendezvous channels when two nodes' clocks are asynchronous, which increases the chances of rendezvous failure and, in turn, link establishment failure. A CH protocol needs to satisfy two critical requirements: (1) establish pairwise rendezvous between two CH sequences of the sender and the receiver on *multiple* channels or *every* available channel in the ideal case; and (2) ensure that the two nodes' CH sequences achieve rendezvous with an *upper bounded* time-to-rendezvous (TTR), even if their clocks are *asynchronous*. The first requirement ensures that the rendezvous failures due to the presence of primary user signals are reduced or minimized. The second requirement makes it possible to upper bound the medium access latency of secondary user nodes since the time required for rendezvous is often unpredictable when establishing links in distributed networks.

We coin the term *rendezvous problem* in the context of time-synchronous or time-asynchronous CR networks to denote the following problem: *How can two channel hopping secondary nodes, with or without clock synchronization, achieve pairwise rendezvous with a guaranteed latency upperbound in the presence of primary user transmissions?*

1.2.2 Spectrum Sharing

In infrastructure-based CR networks, i.e., the 802.22 wireless regional area networks (a.k.a. WRANs), coexisting networks have to share the fallow spectrum to avoid co-channel interference for satisfying the quality of service (QoS) requirements of their admitted service workloads in an OSS environment.

IEEE 802.22 specifies the air interface for a wireless regional area network (WRAN) that operates in fallow TV broadcast bands [50]. An 802.22 *cell* is a single-hop, point-to-multipoint wireless network composed of a Base Station (BS) and several Consumer Premise Equipments (CPEs). Incumbent services refer to TV broadcasting services or services for Part 74 devices[2] (wireless microphones) operating in TV bands, and secondary users refer to IEEE 802.22 entities (BS and CPEs).

802.22 prescribes two types of inter-BS dynamic resource sharing mechanisms so that cells can satisfy the QoS of their admitted workloads. A BS in need of spectrum (*renter* BS) is allowed to selectively rent *candidate* channels of neighboring cells (surplus channels that can be given up) via the Resource Renting mechanism. Note that a candidate channel of one cell may be an *active* channel (a channel in use) of another cell. Thus, Resource Renting is a *non-exclusive spectrum sharing* scheme that enables neighboring BSs to share one or more channels. 802.22 also defines an *exclusive spectrum sharing* scheme called inter-BS *On-Demand Spectrum Contention (ODSC)*. In ODSC, a BS in need of spectrum (*contention source*) selectively contends for candidate channels of neighboring BSs (*contention destinations*). If the contention source wins the contention, it occupies the contended channels exclusively, while the contention destinations vacate those channels via channel switching.

Unfortunately, 802.22's inter-BS resource sharing mechanisms do not consider incumbent coexistence issues when a renter BS selects channels to rent. If the appearance of incumbent signals in the rented channels is frequent, it is likely to have two negative impacts: the QoS degradation due to frequent channel switches in the renter cell when incumbent signals are detected by the renter and interference experienced by incumbent users. Another shortcoming of the resource sharing mechanisms is their failure to adequately address self-coexistence issues. The non-exclusive spectrum sharing scheme does little to prevent self-interference among co-channel overlapping cells, which can render 802.22 networks to be useless [22]. Although the exclusive spectrum sharing scheme can avoid self-interference altogether, it incurs heavy control overhead due to its channel contention procedure.

Therefore, an appropriate tradeoff has to be made between self-interference and control overhead. An ideal spectrum sharing protocol needs to have the ability to dynamically "switch" between non-exclusive and exclusive spectrum sharing when the situation requires it.

1.2.3 Frequency Reuse of a Single Channel

In a densely populated urban area, co-channel spectrum sharing is needed to accommodate the demands of co-located infrastructure-based CR networks operating in limited spectrum. In particular, a new problem arises—the forced sharing of a single white-space channel among multiple uncoordinated CR network cells managed by different network operators. Each operator manages a network consisting of

---

[2] Part 74 devices are low-power wireless devices, such as wireless microphones, which are licensed to operate in the TV broadcast bands.

multiple cells, and coordinates the cells to avoid harmful inter-cell interference. We are especially interested in the worst case where each co-located network cell in the area of interest belongs to a different operator. In such case, global network deployment and frequency planning are not available, and co-channel spectrum sharing without inter-cell coordination is challenging.

We recognize that wireless channels can have different physical-layer characteristics, such as fading, shadowing, and noise [34]. However, we focus on the sharing of a single channel and are particularly interested in the impact of inter-cell interference (in the category of noise) on resource allocation. Other characteristics of shared narrowband subchannels are out of the scope of this study. The shared channel is a white-space (or unlicensed) channel, which is free of incumbent signals and is locally available for all the cells in the area of interest. These co-located cells are considered as co-channel cells, in which active sessions share the same available channel.

1.2.4 Channel Assignment in Multi-hop Cognitive Radio Networks

In the past few years, several channel assignment strategies have been proposed for conventional multichannel wireless networks.

- In link-based channel assignment, channel assignment is performed at the granularity of a link between two given nodes. All packets on this link are transmitted on the same channel as long as the channel assignment does not change.
- The flow-based approach assigns channels at the granularity of a flow, i.e., packets of a flow are scheduled on the same channel.
- In [105], Vedantham et al. proposed component-based channel assignment. (See [105] for a precise definition of a component). In this approach, nodes belonging to a set of intersecting flows (defined as a "component") are assigned to the same channel. Vedantham et al. showed that the component-based approach's theoretical performance does not lag significantly behind finer granularity channel assignment (i.e., link- and flow-based) approaches when channel switching delay is ignored.

It is not possible to apply component-based/flow-based channel assignment to CR networks because not all of the secondary nodes within a given component/flow have access to the same set of channels due to the temporal and spatial spectrum variability caused by primary user's spectrum utilization. Hence, it is no surprise that existing channel assignment schemes for CR networks use the link-based approach [13, 15, 120, 121]. Unfortunately, if channel switching delay is significant, it is also showed in [105] that the component-based channel assignment strategy has practical advantages over existing strategies, such as the avoidance of channel switching delay and overhead. The link-based approach suffers from these practical limitations because a single radio node serving two links assigned with different channels has to switch between channels when forwarding data packets, and thus degrading the overall network performance.

In this study, we consider the channel assignment problem in a single radio interface, multi-hop CR network, and a new channel assignment strategy is needed to handle the spectrum variability as well as minimize the demand for channel switching by using an appropriate granularity of channel assignment.

1.2.5 Coexistence of Heterogeneous Cognitive Radio Networks

The coexistence of CR networks becomes more challenging when the coexisting networks are heterogeneous (i.e., use different air interface standards or radio access technologies). As dynamic spectrum access (DSA) technologies evolve, we can expect that white space spectrum is becoming crowded with various wireless systems, as what happened to the Industrial, Scientific, and Medical (ISM) bands.

As the most representative white space spectrum, TVWSs have received great attention from the wireless community due to the desirable signal propagation characteristics in the TV bands and the overcrowding in the ISM bands. A number of emerging standards prescribe wireless access technologies operating in the TVWSs, and more standards are expected to follow suit. These heterogeneous secondary wireless access technologies have equal priority, and are expected to coexist in the same TV bands. Recently, the FCC is opening up 3.5 GHz band for new DSA opportunities. The three-tier spectrum sharing framework in the 3.5 GHz band requires heterogeneous database-driven wireless access technologies that may not have equal priority to coexist safely. In these types of white space spectrum for DSA, the coexistence of various wireless networks necessitates the need for *heterogeneous coexistence* (HC) mechanisms [8, 37] to mitigate harmful mutual interference among the coexisting networks.

Different from HC issues in the ISM bands, it is more challenging to support HC in OSS due to the signal propagation characteristics in licensed bands, the spectrum agility of systems operating in white space spectrum, and the disparity of PHY/MAC strategies of the coexisting systems. The excellent RF penetration depth of signals in TV bands can cause high level of inter-network interference, while the federal incumbents in 3.5 GHz band are very sensitive to harmful interference from lower-tier users. Furthermore, the heterogeneity of coexisting networks in terms of network architecture, radio access technologies (RATs), and required QoS poses very challenging problems in the design of MAC-layer HC mechanisms.

Chapter 2
Taxonomy of Coexistence Mechanisms

In the pervious chapter, we identify a few problems that are related to the medium access control (MAC) layer protocol design for coexistence of CR networks. To offer a clear picture of coexistence issues and related technical challenges to these problems, in this chapter, we propose a taxonomy of coexistence mechanisms for CR networks [31]. These background knowledge will facilitate the understanding of the various nomenclature and concepts that will be used throughout the book.

We first briefly review recently published or emerging wireless standards that prescribe license-exempt operation in TVWSs, including those for heterogeneous coexistence [8, 37] that can be applied to homogeneous coexistence scenarios. In December 2009, ECMA-392 [24] was finalized as the first standard for personal/portable CR devices operating in TVWSs. It specifies MAC and PHY operations and defines several self-coexistence mechanisms for inter-network coordination and interference mitigation. In 2008, Google and Microsoft proposed the idea of WiFi-like operation in TVWSs, called WiFi 2.0 or WhiteFi. A new standard based on this idea was formalized as IEEE 802.11af [53], which targets higher rate and wider coverage than the current WiFi services by using CR-enabled access points (APs) and user terminals. Besides incumbent protection, IEEE 802.11af also needs to address the coexistence of co-located APs, even though the coexistence mechanisms are yet to be finalized. License-exempt operation of existing licensed networks, e.g., LTE and IEEE 802.16, further creates new challenges. At present, license-exempt LTE [83] is still in its infancy, but IEEE 802.16h [48] was published as a standard amendment for license-exempt WiMAX in July 2010. In IEEE 802.16h, various coordinated and uncoordinated coexistence mechanisms are proposed, which are suitable for the coexistence of metropolitan area networks with heterogeneous others in TVWSs. In July 2011, IEEE 802.22 [50] was released as a new standard for long-range CR networks located in rural areas using TVWSs. Like ECMA-392, several self-coexistence mechanisms are defined in it to mitigate mutual interference among co-located networks belonging to different operators. Furthermore, IEEE 802.19.1 [8] and COGEU project [77] are being developed to provide general solutions to the coexistence of 802 or non-802 networks in various CR-enabled use cases—e.g., campus, apartment complex, and home. A typical IEEE 802.19.1 system consists of a

© Springer International Publishing Switzerland 2014

K. Bian et al., *Cognitive Radio Networks*,

DOI: 10.1007/978-3-319-07329-3_2

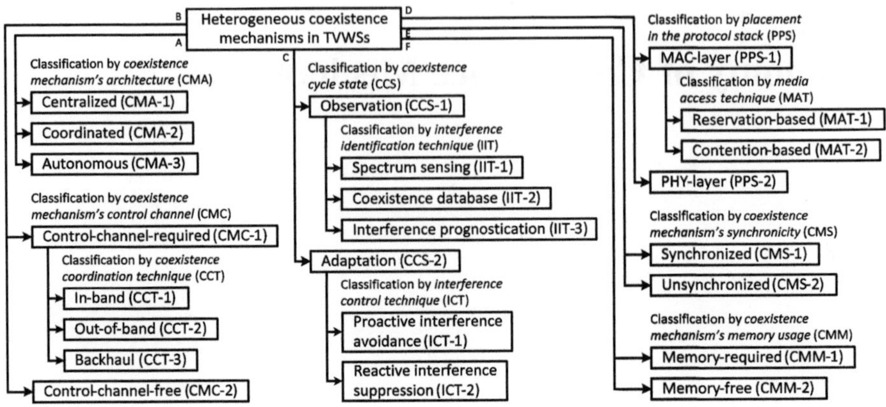

Fig. 2.1 A taxonomy of coexistence mechanisms in TVWSs

coexistence manager, which acts as a centralized resource allocator, and a coexistence enabler, which acts as a coexistence information collector maintaining interfaces between the coexistence enabler and coexisting CR networks. The HC issues are common not only in TVWSs but also in ISM bands. A widely studied coexistence scenario is the co-channel coexistence of low-power IEEE 802.15 networks and high-power IEEE 802.11 networks [81–118]. This problem is also addressed by IEEE 802.15.2. Another popular coexistence scenario is the co-channel coexistence of contention-based IEEE 802.11/802.15 networks and reservation-based IEEE 802.16 networks [57, 71]. The coexistence mechanisms defined in these standards will be introduced as examples in detail.

The proposed taxonomy classifies coexistence mechanisms using a diverse set of criteria as shown in Fig. 2.1. For each category of the taxonomy, we provide examples and discuss the pros and cons of related coexistence mechanisms. In Table 2.1, we map coexistence mechanisms, which have been proposed in the literature or standards, to our taxonomy.

2.1 Classification by Coexistence Mechanism's Architecture (A)

The sharing of TVWSs among coexisting networks can be achieved in several ways depending on whether or not decision-making coexistence infrastructures and inter-network coordination channels are available. Based on the coexistence mechanism's architecture (CMA), coexistence mechanisms are classified into centralized, coordinated, and autonomous categories.

CMA-1: Centralized Mechanisms. Centralized mechanisms require both decision-making coexistence infrastructures and inter-network coordination channels. A coexistence scenario for centralized mechanisms is illustrated in Fig. 2.2a. Each

Table 2.1 Mapping of coexistence mechanisms to classification methods

	Adaptive modulation and coding	Centralized coexistence framework	Coexistence beacon signaling	Coexistence control channel	Coexistence frame scheduling	Coexistence information database	Cognitive pilot channel	Collocated coexistence messaging	Cooperative busy tone signaling	Coordinated contention-based protocol	Credit-token-based coexistence protocol	Dynamic frequency/channel selection	Interference cancellation and suppression	Internet-server-facilitated messaging	Listen before talk	Opportunistic channel access	Smart antenna	Time/frequency-division multiple access	Transmit power control
CMM-2	✓		✓	✓	✓			✓	✓	✓	✓				✓		✓		✓
CMM-1		✓				✓	✓						✓	✓	✓		✓		✓
CMS-2	✓									✓					✓		✓	✓	✓
CMS-1		✓	✓	✓	✓	✓	✓	✓			✓	✓		✓	✓			✓	
PPS-2	✓												✓				✓		✓
PPS-1: MAT-2									✓	✓					✓	✓			
PPS-1: MAT-1		✓	✓	✓	✓	✓	✓			✓	✓			✓			✓		
CCS-2: ICT-2	✓												✓			✓		✓	
CCS-2: ICT-1		✓			✓					✓	✓	✓			✓		✓		
CCS-1: IIT-3																✓			
CCS-1: IIT-2						✓													
CCS-1: IIT-1			✓	✓	✓		✓	✓	✓	✓	✓	✓		✓	✓				
CMC-2	✓										✓	✓		✓	✓	✓		✓	
CMC-1: CCT-3		✓			✓									✓			✓		
CMC-1: CCT-2				✓	✓		✓			✓	✓								
CMC-1: CCT-1			✓					✓	✓										
CMA-3	✓										✓	✓			✓	✓	✓	✓	
CMA-2		✓	✓	✓	✓	✓	✓	✓	✓	✓	✓		✓			✓			
CMA-1	✓																	✓	

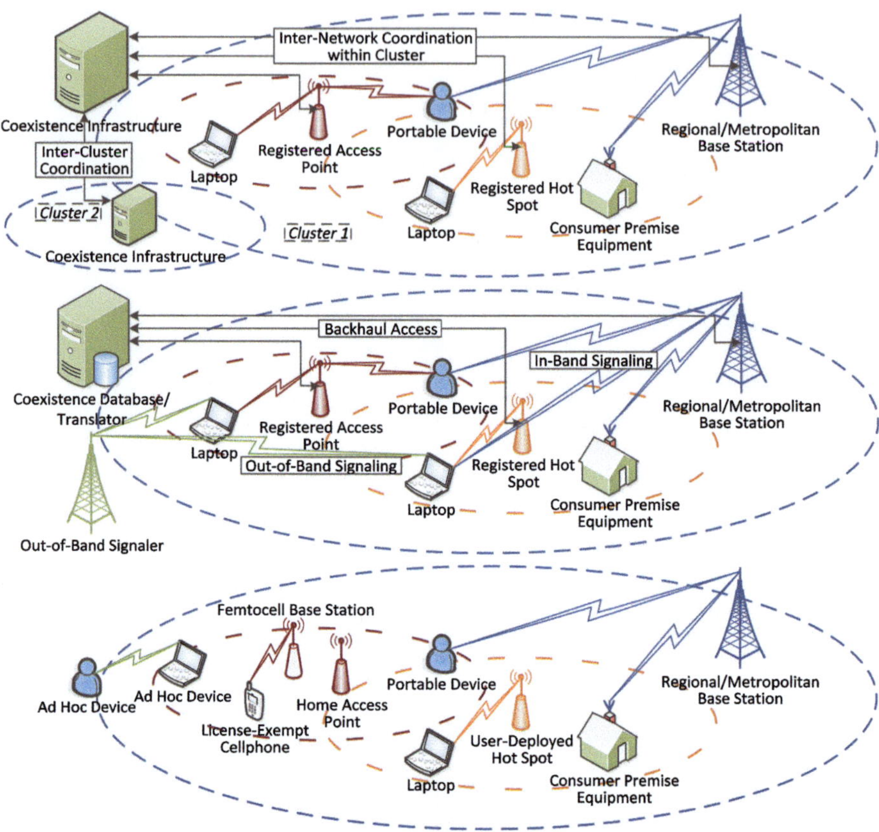

Fig. 2.2 Examples of coexistence scenarios: **a** centralized mechanisms (*top*); **b** Coordinated mechanisms (*middle*); and **c** Autonomous mechanisms (*bottom*)

operator can deploy a single or multiple infrastructures to carry out centralized spectrum sharing via a star-topology or a cluster-based architecture. In each cluster, as in Fig. 2.2a, each coexistence infrastructure minimizes interference among registered coexisting networks based on centrally collected coexistence information. Furthermore, multiple coexistence infrastructures can coordinate with each other for further collaboration. For example, the standard-independent centralized coexistence framework in IEEE 802.19.1 defines a coexistence manager for central management of all associated networks. In addition, the collaboration of multiple 802.19.1 coexistence managers or cluster-head equipments [107] can be supported through certain inter-cluster coordination interfaces. Centralized mechanisms can operate independently of existing wireless standards, and thus, registered networks can be heterogeneous and no major standard modifications are required. They can effectively minimize inter-network interference by utilizing centrally collected global coexistence information. However, centralized solutions incur costly new infrastructures and subsequent operational costs. The effectiveness of centralized mechanisms is diminished when a significant fraction of the coexisting devices/networks are not registered and not under central control.

CMA-2: Coordinated Mechanisms. Coordinated mechanisms are applied when each coexisting network locally conducts resource allocation without the need for extra decision-making coexistence infrastructures but inter-network coordination channels are still available. A coexistence scenario that employs coordinated mechanisms is illustrated in Fig. 2.2b. The coexisting networks coordinate with each other through various information signaling and retrieving techniques. Based on the collected coexistence information, each network can make decisions to mitigate inter-network interference. For example, the cooperative busy tone signaling technique [118] helps high-power IEEE 802.11 networks detect signaling messages from co-channel, low-power IEEE 802.15.4 networks to prevent 802.11 devices from dominating the channel contention process. Coordinated mechanisms achieve effective inter-network interference mitigation. However, they are often limited to certain coexistence scenarios. More details about coordination channels will be discussed in the next subsection.

CMA-3: Autonomous Mechanisms. When neither decision-making coexistence infrastructures nor inter-network coordination channels are available, each coexisting network has to utilize autonomous mechanisms to achieve best-effort interference mitigation. A coexistence scenario for autonomous mechanisms is illustrated in Fig. 2.2c. Each coexisting network performs resource allocation and manages inter-network interference only based on local observation. For example, the dynamic frequency/channel selection technique enables each network to select or switch to the channel with the least amount of interference based on the local evaluation of channel quality. The listen before talk policy prescribes a device to access spectrum based on the outcome of local spectrum sensing. Autonomous mechanisms are low in complexity and can adapt to dynamic environments. They can be integrated with centralized and coordinated mechanisms to build hybrid mechanisms. However, autonomous mechanisms by themselves may not sufficiently mitigate inter-network interference due to their best-effort nature.

2.2 Classification by Coexistence Mechanism's Control Channel (B)

The availability of control channels for inter-network coordination directly determines the design of coexistence mechanisms. Based on the coexistence mechanism's control channel (CMC), coexistence mechanisms are classified into control channel-required and control channel-free categories.

CMC-1: Control Channel-Required Mechanisms. Both centralized mechanisms (CMA-1) and coordinated mechanisms (CMA-2), whose operations necessarily require inter-network coordination, fall into this classification. Based on the coexistence coordination technique (CCT), control channel-required mechanisms are further classified into in-band, out-of-band, and backhaul categories.

CMC-1: CCT-1: In-Band Mechanisms. To deliver coexistence information to coexisting neighbors, each device can periodically broadcast coexistence signaling messages on its data channels, such as in-band signaling illustrated in Fig. 2.2b. For example, the coexistence beacon signaling in ECMA-392 and IEEE 802.22 defines specific time slots in each network's regular superframes for periodic broadcast of coexistence beacons. Although such a technique is used for self-coexistence purpose here, it can be a coexistence mechanism when coexisting networks can decode signaling messages between them. The co-located coexistence messaging technique [29] lets each multi-radio user terminal forward in-band signaling messages among its multiple heterogeneous home networks. In-band mechanisms do not require extra infrastructures or control channels for inter-network coordination. However, they only work when coexisting networks use the same radio access technology or when heterogeneous devices can decode each others' signaling messages.

CMC-1: CCT-2: Out-of-Band Mechanisms. Instead of relying on data channels, coexisting networks can broadcast coexistence signaling messages on a dedicated or dynamically established common control channel, such as out-of-band signaling illustrated in Fig. 2.2b. For example, the coexistence control channel in IEEE 802.16h supports secondary synchronization, user detection, interference evaluation, and inter-system communication. The cognitive pilot channel [37] always carries up-to-date coexistence information, broadcasted by operators or third-party entities, which can be retrieved by each network on demand. Out-of-band mechanisms can be used in direct inter-network negotiations if a standardized signaling message format is adopted. However, they fully rely on the existence and reliability of a common control channel.

CMC-1: CCT-3: Backhaul Mechanisms. When wired backhaul links are available, these links can be used to coordinate spectrum access among the coexisting networks. As shown in Fig. 2.2b, each network can either access a central coexistence database or coordinate with neighbors using the method of message translation. For example, the coexistence information database [8], which is implemented on servers connected to the Internet, provides on-demand lookup functionality. The Internet server-facilitated messaging technique [8] enables inter-network coordination over the Internet by providing message translation and forwarding services. Backhaul mechanisms can provide relatively complete knowledge of the spectrum environment in the vicinity of each network in a short time. They enable active reporting and retrieving of coexistence information instead of passive network detection. However, backhaul links may not cover all the coexisting networks to make coexistence information incomplete.

CMC-2: Control Channel-Free Mechanisms. In the context of our taxonomy, control channel-free mechanisms are classified under a category that is equivalent to the one that autonomous mechanisms (CMA-3) are classified under.

2.3 Classification by Coexistence Cycle State (C)

The simplified cognition cycle of a CR consists of three states: observation, decision, and adaptation [73]. Based on this cycle, we propose a classification of coexistence mechanisms based on the coexistence cycle state (CCS), which includes: observation and adaptation categories.

CCS-1: Observation Mechanisms. The purpose of observation state is to identify the presence or even the source of inter-network interference. Based on the interference identification technique (IIT), observation mechanisms are further classified into spectrum sensing, coexistence database, and interference prognostication categories.

CCS-1: IIT-1: Spectrum Sensing Mechanisms. For the detection of coexisting networks, each CR network can either locally scan TVWSs via spectrum sensing or cooperatively exchange coexistence information with neighbors through inter-network coordination channels. Note that the purpose of spectrum sensing here is to detect coexisting secondary networks instead of primary incumbents. For example, the scanning-based opportunistic channel access technique [81] enables each IEEE 802.15.4 network to scan multiple channels potentially interfered by IEEE 802.11 networks, and keep searching for a better channel selection by using simulated annealing optimization method. The metric of channel quality is computed in terms of the energy of detected 802.11 interference and the number of heard 802.15.4 beacons. The coexistence beacon signaling and coexistence control channel techniques facilitate cooperative spectrum sensing and coexistence information exchange by providing means for in-band or out-of-band inter-network coordination. Spectrum sensing mechanisms are relatively easy to implement. However, they are unreliable in real-world coexistence scenarios, and need to be augmented with other methods for reliable detection of coexisting networks.

CCS-1: IIT-2: Coexistence Database Mechanisms. A coexistence database storing geolocation and operation information about secondary CR networks can be utilized by each coexisting network to help identify potential sources of interference. For example, the coexistence information database in IEEE 802.16h stores the shared information regarding the actual and intended usage of spectrum resource for certain local regions. Besides base stations, user terminals also contribute to complete the database by providing interference information pertinent to themselves. Coexistence database mechanisms provide a more practical and effective means of detecting coexisting networks. However, they cannot detect the presence of unregistered devices/networks at the database authority.

CCS-1: IIT-3: Interference Prognostication Mechanisms. The past spectrum sensing results can help each network make predictions on the availability of spectrum and potential interference. Techniques for interference prognostication include modeling and machine learning. These techniques can be used to predict the behavior of potential interferers by leveraging their past spectrum access behavior. For example, the learning-based opportunistic channel access techniques [46, 81] enable each IEEE 802.15.4 network to learn the statistical regularity of IEEE 802.11 operation and predict the opportunities of white spaces free of 802.11 traffic. The ideas are similar to the modeling of primary users' behavior in order to empirically prevent secondary users from using the same spectrum being occupied by primary users. Interference prognostication mechanisms offer low-power networks more opportunities of spectrum access by circumventing high-power networks. However, their effectiveness highly relies on the accuracy of predictions, which is very challenging to guarantee for the operation in TVWSs.

CCS-2: Adaptation Mechanisms. Once inter-network interference has been detected, a network needs to adapt to its interference environment by taking various measures to change its transmission characteristics. Based on the interference control technique (ICT), adaptation mechanisms are further classified into proactive interference avoidance and reactive interference suppression categories.

CCS-2: ICT-1: Proactive Interference Avoidance Mechanisms. Upon detecting interference from neighbors, each CR network can choose to directly switch to a better-quality channel, if such a channel is available, or wait until the channel becomes vacant. If coordination channels are available, inter-network interference can even be avoided in advance. For example, the dynamic frequency/channel selection technique provides a certain network with a list of candidate channels that can be used for channel switching whenever needed. The time/frequency-division multiple access is a common interference avoidance technique that centrally enables coexisting networks to operate in separate time slots or channels. Proactive interference avoidance mechanisms are suitable for the coexistence of CR networks, since CR devices are capable of dynamic spectrum access. However, they necessarily require the support of observation mechanisms to identify the candidate channels free of interference.

CCS-2: ICT-2: Reactive Interference Suppression Mechanisms. Interference suppression techniques are used to alleviate or suppress interference but do not enable a network to avoid it. For example, the interference cancellation and suppression technique [71] enables a network that is experiencing interference to utilize adaptive filters or prior knowledge of interferers to estimate and cancel the interference step by step. The transmit power control policy can be used to mitigate co-channel or adjacent-channel interference when the coexisting networks are managed by the same operator. Reactive interference suppression mechanisms help to support non-exclusive co-channel spectrum sharing as long as inter-network interference is tolerable. However, their effectiveness is usually limited, especially when interference comes from high-power interferers that belong to different operators.

2.4 Classification by Placement in the Protocol Stack (D)

In general, inter-network interference is mitigated in the MAC or PHY layer of the protocol stack. Based on the placement in the protocol stack (PPS), coexistence mechanisms are classified into MAC-layer and PHY-layer categories.

PPS-1: MAC-Layer Mechanisms. Most coexistence mechanisms are placed in the MAC layer. Medium access can be performed in one of two ways—each network can choose to either reserve or contend for spectrum resource. Based on the media access technique (MAT), MAC-layer mechanisms are further classified into reservation-based and contention-based categories.

PPS-1: MAT-1: Reservation-Based Mechanisms. Reservation-based spectrum sharing can be achieved in the time, frequency, and space domains (space-division mechanisms are included in the PPS-2 category). In the time domain, multiple co-channel networks can take turns to access the shared channel in separate time frames or slots. For example, the coexistence frame scheduling in IEEE 802.16h divides time frames into Master, Slave, and Shared subframes, which can be scheduled by each base station for uplink and downlink in a flexible mode. The operation of Master systems in their Master subframes should be protected from harmful interference caused by concurrent Slave systems, and coexisting networks equally share the role of Master system on a rotating basis. In Shared subframes, all the coexisting networks may operate in parallel under the limits on transmit power levels. In the frequency domain, multiple coexisting networks can simultaneously access the same TVWSs but use separate channels or sub-channels by direct or orthogonal spectrum splitting. Furthermore, time-frequency resource blocks may be conceived for greater flexibility and granularity in spectrum sharing. For example, the credit token-based coexistence protocol in IEEE 802.16h permits auction-based spectrum leasing among coexisting networks for channel reservation in the subsequent time frames. Each network can be either an offerer or a requester to transfer the "ownership" of spectrum dynamically. Reservation-based mechanisms can guarantee fairness and reliable throughput of coexisting networks regardless of the discrepancy in their channel definitions, signal characteristics, or transmit power levels. However, they need to be supported by inter-network coordination channels or even extra coexistence infrastructures.

PPS-1: MAT-2: Contention-Based Mechanisms. Contention-based media access can also be used to support heterogeneous coexistence. For example, the coordinated contention-based protocol in IEEE 802.16h prescribes networks with reservation-based MAC to periodically halt transmissions so that the resulting idle time frames can be utilized by networks with contention-based MAC. Contention-based mechanisms are easy to implement, and do not require strict inter-network synchronization. However, they do not always guarantee fairness and constantly reliable spectrum access due to the randomness of contention results.

PPS-2: PHY-Layer Mechanisms. In the PHY layer, various techniques can be used to mitigate inter-network interference. For example, the smart antennas can be used to reduce the coexistence-related interference by minimizing side-lobe radiation. The spatial reuse of shared spectrum can be improved by directional interference patterns. The adaptive modulation and coding technique enhances coexistence via dynamically adaptable PHY parameters according to the varying radio environments, such as path loss and interference. Most PHY-layer mechanisms also belong to autonomous mechanisms (CMA-3), so further discussion is neglected.

2.5 Classification by Coexistence Mechanism's Synchronicity (E)

The mitigation of inter-network interference can be facilitated by coexistence mechanisms synchronized across coexisting networks. Based on the coexistence mechanism's synchronicity (CMS), coexistence mechanisms are classified into synchronized and unsynchronized categories.

CMS-1: Synchronized Mechanisms. A number of coexistence techniques require accurate inter-network synchronization, at either the MAC or the PHY layer. For example, the coordinated contention-based protocol requires the coexisting networks with reservation-based MAC synchronize their quiet periods so that the carrier sensing of co-channel networks with contention-based MAC can discover such opportunities. The coexistence beacon signaling and coexistence control channel techniques enable faster detection of coexisting networks if the networks synchronize with each other, due to the periodicity of in-band or out-of-band signaling. Synchronized mechanisms address coexistence issues via precise separation of coexisting networks in the time domain. However, inter-network synchronization is difficult to implement without extra infrastructures to support.

CMS-2: Unsynchronized Mechanisms. Inter-network synchronization is not necessary for a number of coexistence techniques. For example, the listen before talk policy enables coexisting networks to contend for spectrum access in an asynchronous manner. The cooperative busy tone signaling technique is proposed as an enhancement of CSMA protocol for IEEE 802.15.4 networks, and does not require synchronization with co-channel IEEE 802.11 networks using CSMA protocol as well. Unsynchronized mechanisms can be readily implemented. However, in most cases, they can only alleviate inter-network interference but cannot avoid it.

2.6 Classification by Coexistence Mechanism's Memory Usage (F)

Certain coexistence mechanisms require the storage of coexistence information. Based on the coexistence mechanism's memory usage (CMM), coexistence mechanisms are classified into memory-required and memory-free categories.

CMM-1: Memory-Required Mechanisms. In some coexistence techniques, memory is needed to store necessary coexistence information. For example, the coexistence information database stores up-to-date geolocation and operation information about secondary CR networks. The machine learning-based opportunistic channel access needs to record recent history of spectrum sensing results and maintain a knowledge base to make reasonable interference predictions. Memory-required mechanisms can help make faster and more thoughtful decisions. However, the costs for memory consumption can be high especially in large-scale complex coexistence scenarios, e.g., an apartment building in a dense urban area.

CMM-2: Memory-Free Mechanisms. A number of coexistence techniques do not require memory usage or only require a negligible size of memory. For example, the co-located coexistence messaging technique for each multi-radio user terminal directly forwards signaling messages from one of its home networks to another one without the need for recording the messages. The cooperative busy tone signaling technique only requires IEEE 802.15.4 signalers to emit busy tones to IEEE 802.11 receivers for spectrum reservation. The listen before talk policy is another typical example. Memory-free mechanisms are suitable for autonomous networks that do not have much system resource. However, their achieved performance in terms of fairness, spectrum utilization, or throughput may not be as good as that by memory-required mechanisms due to limited coexistence knowledge.

Chapter 3
Rendezvous of Cognitive Radios

In Chap. 1, we introduce the rendezvous/control channel establishment problem in the MAC layer protocol and the design requirements for a channel hopping (CH) based rendezvous establishment protocol, a.k.a. CH schemes. However, most, if not all, of the existing CH schemes [6, 23, 91, 93, 103] only provide ad-hoc approaches for generating CH sequences and evaluating their properties. In a time-synchronous CR network, we present a systematic approach in this chapter, based on *quorum systems*, for designing and analyzing CH protocols for the purpose of control channel establishment. The proposed approach, called *Quorum-based Channel Hopping* (*QCH*) system, utilizes the *intersection property* of quorum systems to generate CH sequences that enable rendezvous on multiple channels between any two CH sequences. Under the assumption of global clock synchronization, we describe two QCH systems. The first system minimizes the upperbound of the time-to-rendezvous (TTR) value between any two CH sequences. The second system evenly distributes the rendezvous points over different timeslots during a CH period, thereby alleviating the *rendezvous convergence* problem. To address the rendezvous problem in time-asynchronous CR networks, we propose three asynchronous channel hopping (ACH) systems that do not require global clock synchronization. The first ACH system is constructed using the cyclic quorum system, the second ACH system is constructed using the array-based quorum system, and the third ACH system is built by leveraging bit-sequence design techniques. Note that, CH sequences in an ACH system are generated by exploiting the *rotation closure property* of quorum systems, which enables the TTR between any two nodes' CH sequences to be upper bounded without requiring clock synchronization.

In this chapter, we first review existing MAC layer protocols with the purpose of rendezvous/control channel establishment in Sect. 3.1. In Sects. 3.3 and 3.4, we describe the system model and the QCH system, respectively. We present two optimal synchronous QCH systems in Sect. 3.5. In Sect. 3.5.3, we compare our proposed QCH systems and existing CH schemes using simulation results. We define the asynchronous channel hopping System in Sect. 3.6. In Sects. 3.6.1 and 3.6.2, we describe two asymmetric ACH systems that are constructed by cyclic quorum systems and array-based quorum systems, respectively. We formulate the optimal ACH system design

© Springer International Publishing Switzerland 2014

K. Bian et al., *Cognitive Radio Networks*,

DOI: 10.1007/978-3-319-07329-3_3

problem in Sect. 3.6.3. In Sects. 3.6.4 and 3.6.5, we present an asymmetric optimal ACH system and a symmetric optimal ACH system. We analytically compare the proposed ACH system designs with existing CH schemes in Sect. 3.6.6. We evaluate our proposed ACH schemes using simulation results in Sect. 3.6.7, and summarize this chapter in Sect. 3.7.

3.1 Rendezvous-Enabling Techniques

3.1.1 Common Control Channel Based Rendezvous

In a typical multi-channel MAC protocol, nodes contend for channels by exchanging MAC layer control packets in the common control channel (CCC) [92, 113]. If there is no access point or base station, channel negotiation is carried out in a distributed manner, between each pair of sender and the intended receiver. We describe CCC-based rendezvous techniques using the example shown in Fig. 3.1. Since there is no existing standard that defines such a MAC layer, the discussions given here are based on the common features shared by most of the MAC protocols using a global or local common control channel [19, 54, 78, 92, 120]. During channel negotiation phase, the following types of MAC frames are utilized: Free Channel List (FCL), SELection (SEL), and REServation (RES). Figure 3.1 shows a simple example of a channel negotiation process between a sender and a receiver. Here, the "sender" is the host that transmits the MAC data frames, and the "receiver" is the host that receives the MAC data frames. The sender first identifies fallow spectrum bands and maps them into logical channels. Then it obtains a free channel list (FCL frame) and sends this to the receiver after waiting a random back-off time. Upon receiving the FCL, the receiver identifies available data channels common to both sides, and then selects one data channel according to a data channel selection policy. The channel selection is indicated in a SEL frame and sent back to the sender. After receiving the SEL frame, the sender notifies its neighbors of the channel selection via a channel reservation message (RES frame). Neighbor nodes refrain from transmitting by maintaining a network allocation vector (NAV) specified in the FCL and SEL frames overheard during the channel negotiation process. Using the above process, a sender and a receiver select a data channel for communicating.

A common control channel will cause two problems in multi-channel wireless networks. First, relying on a common control channel may create a bottleneck (a la *control channel saturation problem*) for the network performance [6, 92]. The second problem that may arise from the use of a common control channel is the single point of failure problem when the control channel is targeted by jamming attacks. In [62], channel hopping is used for recovering the control channel for a cluster of ad hoc nodes whenever the common control channel is jammed. The hopping sequence of the control channel is pseudo random, but it is pre-determined. The hopping sequence is distributed by the cluster head to all nodes in the same cluster. In [21], a dynamic

Fig. 3.1 Distributed channel negotiation process between a sender and a receiver

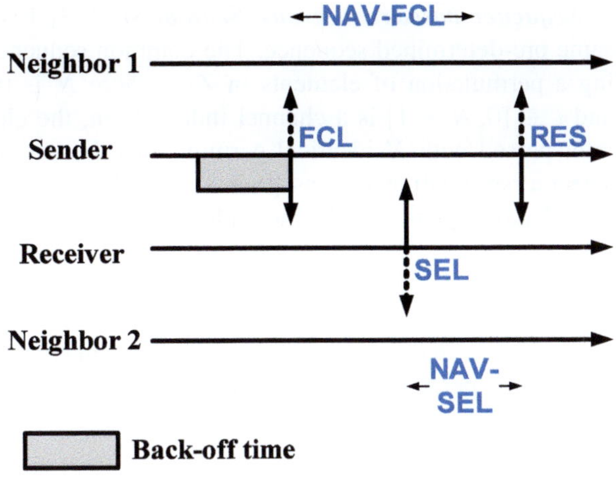

common control channel scheme is proposed, in which the common control channel can switch to other channels if the current spectrum band is occupied by the incumbent users. However, control channel hopping techniques cannot solve the control channel saturation problem.

3.1.2 Channel Hopping Based Rendezvous

In the design of multi-channel MAC protocols, the use of channel hopping (or parallel rendezvous) techniques have been proposed to enable rendezvous between a pair of communicating nodes, and the CH sequences are carefully generated to guarantee that different CH sequences must have overlap.

Blind rendezvous (BR) channel hopping [91]. In this scheme, each node hops from one channel to another randomly. At a particular instant, a node occupies one of these channels with probability $1/N$, where N is the total number of channels. When two nodes occupy the same channel at the same time, rendezvous occurs. The BR scheme does not guarantee a bounded TTR between any two sequences.

Slotted Seeded Channel Hopping (SSCH) [6]. Each node is allowed to have one or multiple (channel, seed)-pairs to determine its CH sequences. SSCH allows $(N-1)$ seeds, where N is the total number of channels. Each sequence period includes a parity slot at which time instant all nodes with the same seed are guaranteed to rendezvous on a channel indicated by the seed value. When each node selects one (channel, seed)-pair, the resulting sequence period is $(N+1)$ timeslots, and each pair of sequences rendezvous exactly once within a period. By design, SSCH is a synchronous CH system (the boundaries of nodes' CH sequences should be aligned), although results in [6] show that it can tolerate moderate clock skew. The amount of clock skew used in [6] to evaluate SSCH is very small relative to one slot duration.

Sequence-based rendezvous (SeqR or SR) [23]. Every node using SR follows the same pre-determined sequence. The common sequence, u^*, is built by first selecting a permutation of elements in $\mathbf{Z}_N$, where N is the total number of channels and $i \in [0, N-1]$ is a channel index. Then, the elements of the permutation is interspersed with N identical permutations. For example, when $N = 3$, one can select a permutation such as $\{0, 2, 1\}$, and the resulting CH sequence u^* would be an ordered sequence of channel indexes, such as

$$\{0, 0, 2, 1, 2, 0, 2, 1, 1, 0, 2, 1\}.$$

The elements in the permutation $\{0, 2, 1\}$ is interspersed with the three replications of the same permutation. The sequence period is $N(N + 1)$ timeslots.

3.2 Quorum Systems

In this chapter, the quorum system will be used for designing CH sequences in a systematic way.

Definition 1 Given a finite universal set $U = \{0, \ldots, n-1\}$ of n elements, a quorum system S under U is a collection of non-empty subsets of U, which satisfies the intersection property:

$$p \cap q \neq \emptyset, \forall p, q \in S.$$

Each $p \in S$ (which is a subset of U) is called a *quorum*.

The quorum systems have been used for designing the wake-up schedule in wireless sensor or ad hoc networks for power saving [56, 102, 122]. The intersection property of quorum systems is exploited to create the overlap between wake-up schedules of two sensor nodes so that they could discover each other without keeping awake all the time. As a result, the power consumption is reduced. In this chapter, we use the intersection property of quorum systems to create overlap between CH sequences when their boundaries are synchronized. When the clock synchronization is unavailable, a second property of quorum systems—rotation closure property—will be used to maintain the overlap between *asynchronous* CH sequences.

Cyclic quorum systems. Here, we provide some definitions related to cyclic quorum systems since those systems are utilized to design channel hopping schemes in Sects. 3.4 and 3.6. The *cyclic quorum system*, first introduced in [68], can be constructed using cyclic difference sets in combinatorial theory [94].

Definition 2 A set $D = \{a_1, a_2, \ldots, a_\kappa\} \subset \mathbf{Z}_n$ is called a relaxed cyclic (n, κ)-difference set if for every $d \not\equiv 0 \bmod n$ there exists at least one ordered pair (a_i, a_j), where $a_i, a_j \in D$, such that $a_i - a_j \equiv d \bmod n$. Here, $\mathbf{Z}_n$ denotes the set of nonnegative integers less than n.

Definition 3 A group of sets

$$B_i = \{a_1 + i, a_2 + i, \ldots, a_\kappa + i\} \mod n, i \in \{0, 1, \ldots, n-1\}$$

is a cyclic quorum system if and only if $D = \{a_1, a_2, \ldots, a_\kappa\}$ is a relaxed cyclic (n, κ)-difference set.

For example, $D = \{0, 1, 3\}$ is a relaxed cyclic $(7, 3)$-difference set under $\mathbf{Z}_7$ since each $d \in \{1, .., 6\}$ is congruent to the difference of two elements in D. Given D,

$$S = \{B_0, B_1, \ldots, B_6\}$$

is a cyclic quorum system under $\mathbf{Z}_7$, where

$$B_i = \{0 + i, 1 + i, 3 + i\} \mod 7, i = \{0, 1, \ldots, 6\}.$$

It was proven in [56] that any quorum q in a cyclic quorum system under $U = \{0, \ldots, n-1\}$ must have a cardinality $|q| \geq \sqrt{n}$.

Given any n, a difference set as small as $\kappa \approx \sqrt{n}$ can be found when $\kappa^2 - \kappa + 1 = n$ and $\kappa - 1$ is a prime power. Such a difference set is called the *Singer* difference set [20], which is the *minimal* difference set whose size κ approximates the lower bound $\sqrt{n}$. Hence, cyclic quorum systems defined by the Singer difference sets are *minimal* cyclic quorum systems in the sense that their quorum sizes are close to the theoretical lower bound. For example, the set $\{1, 2, 4\}$ under $\mathbf{Z}_7$ is a Singer difference set when $\kappa = 3$.

Any set D that contains $\lceil \frac{n+1}{2} \rceil$ elements of $\mathbf{Z}_n$ is a relaxed cyclic $(n, \lceil \frac{n+1}{2} \rceil)$-difference set and a cyclic quorum system $S = \{B_0, B_1, \ldots, B_{n-1}\}$ can be constructed based on D according to Definition 3. Since D contains more than half of the elements in $\mathbf{Z}_n$, we refer to such a cyclic quorum system, S, as a *majority* cyclic quorum system. For example,

$$S = \{\{0, 1, 2\}, \{1, 2, 3\}, \{2, 3, 0\}, \{3, 0, 1\}\}$$

is a majority cyclic quorum system under $\mathbf{Z}_4$.

3.3 Network Model

We assume an OSS environment, where secondary users equipped with CRs dynamically access spectrum. Suppose multiple pairs of (secondary) senders and receivers are operating over N orthogonal frequency channels that are licensed to the primary user. Channels are labeled as $0, 1, \ldots, N-1$. Each node (a node may be a sender or a receiver) is equipped with a single half-duplex transceiver. Hence, a node can only listen to or transmit over one channel at a time. A packet sent over a channel can be heard by any node within the communication range of the transmitting node.

Time-slotted system. We consider a time-slotted communication system, where a global system clock exists. The local clock of each node may be synchronized to the global clock or may differ with the global clock by a certain amount of clock drift. A network node is assumed to be capable of hopping between different channels according to a channel hopping sequence and its local clock. A packet can be exchanged if the sender and the receiver hop onto the same channel at the same timeslot. We assume that one timeslot is long enough to exchange multiple packets.

Channel hopping (CH) sequence. A CH sequence (or a sequence) determines the order with which a node visits all available channels. We represent a CH sequence u of period T as a polynomial of order $(T - 1)$ called the CH Sequence Function (CSF):

$$u = f_u(x) = u_0 + u_1 x + u_2 x^2 + \cdots + u_i x^i + \cdots + u_{T-1} x^{T-1},$$

where $u_i \in [0, N - 1]$ represents the channel index of sequence u in the ith timeslot of a CH period.

Given two sequences of period T, u and v, and their CSFs, $f_u(x)$ and $f_v(x)$, if $\exists i \in [0, T - 1]$ s.t. $u_i = v_i = h$, where $h \in [0, N - 1]$, we say that u and v *rendezvous* in the ith timeslot on channel h. The ith timeslot is called the *rendezvous slot* and channel h is called the *rendezvous channel* between u and v. Let $I_h(u, v)$ denote a function that indicates whether channel h is a rendezvous channel between two sequences u and v, i.e.,

$$I_h(u, v) = \begin{cases} 1, & \text{if } \exists i \in [0, T - 1], \ s.t. \ u_i = v_i = h, \\ 0, & \text{otherwise.} \end{cases}$$

The *number of rendezvous channels* between two sequences u and v, $C(u, v)$, is defined as

$$C(u, v) = \sum_{h=0}^{N-1} I_h(u, v).$$

Channel hopping systems. In the OSS paradigm, the value of $C(u, v)$ directly impacts the *robustness* of the pairwise control channel established using sequences u and v. Recall that secondary users share spectrum opportunistically with incumbent users who have priority access rights. In such a scenario, secondary users are required to vacate the currently occupied channels when incumbent signals are detected in them. This requirement poses a difficult challenge in the design of MAC protocols for CR networks—in particular, in terms of how to establish control channels in such a way that enables the reliable exchange of control information despite the unpredictable appearance of incumbent signals. The robustness of the control channels established using sequences u and v is proportional to the value of $C(u, v)$, since this value determines the number of distinct channels in which the rendezvous occur within a sequence period. If the rendezvous are spread out over a greater number of distinct channels, then the probability of link breakage caused by the inability to exchange control packets (which in turn is due to the appearance of incumbent signals) decreases. Thus, we have the following *channel hopping system design problem*.

Problem 1 Given T, the CH system design problem is to devise a set of CH sequences of period T, denoted as H, which satisfies the following two properties:

1. $\forall u \in H$, u is a polynomial of order $(T-1)$;
2. $m \geq 1$, where $m = \min_{\forall u,v \in H}\{C(u,v)\}$.

The set H is called a *CH system* of period T, and m is *the degree of overlapping* of the CH system H.

It is readily apparent that a CH system H is a quorum system, since it satisfies the intersection property: any two sequences in H have at least one overlap. Each CH sequence in H is a quorum.

Coexistence with incumbent users. All secondary nodes coexist with primary users in the following way: a secondary node is able to detect the primary user signals using spectrum sensing techniques [1], and thus avoids transmitting on a channel where primary user signals are detected. The primary user signals may appear in any channel.

3.4 The Quorum-Based Channel Hopping System

In this section, we introduce an algorithm that uses a quorum system to construct a CH system for any value of $m \in [1, N]$. We refer to this algorithm as Algorithm 1.

Without loss of generality, suppose we want to construct a CH system where every pair of CH sequences rendezvous in m different channels. We randomly select m channels from $\{0, \ldots, N-1\}$ to construct a *set of rendezvous channels*, such as $R = \{h_0, h_1, \ldots, h_{m-1}\}$. In our construction algorithm, every CH sequence is composed of m frames and each frame is composed of k slots (k is called the frame length). Hence, the period of each CH sequence is $T = m \cdot k$. We use the following example to explain the construction algorithm.

Suppose the set of rendezvous channels is $R = \{0, 1, 2\}$, each CH sequence is composed of $m = 3$ frames, and each frame has $k = 3$ slots.

1. First construct a universal set, $U = \mathbf{Z}_k = \{0, 1, 2\}$;
2. Construct a quorum system S under U, $S = \{\{0, 1\}, \{0, 2\}, \{1, 2\}\}$[1];

---

[1] The desired properties of the CH system determines the particular quorum system that is constructed. In Sect. 3.5, we discuss a number of quorum systems that can be used to construct optimal CH systems.

Slot Index:	0	1	2	3	4	5	6	7	8
u	**0**	**0**	2	**1**	**1**	0	**2**	**2**	1
v	**0**	1	**0**	**1**	2	**1**	**2**	0	**2**
w	1	**0**	**0**	2	**1**	**1**	0	**2**	**2**

|←————1st frame————→|←————2nd frame————→|←————3rd frame————→|

Fig. 3.2 An illustration of QCH system Q with $m = 3$ and $k = 3$. Any two sequences overlap on three channels. We use the quorum system $S = \{\{0, 1\}, \{0, 2\}, \{1, 2\}\}$ over $U = \{0, 1, 2\}$ to construct Q. The numbers inside the slots denote channel index values: *bold-font values* denote channel indexes from R and the *non-bold font values* denote channel indexes randomly chosen from $\{0, \ldots, N - 1\}$

3. Using the quorum $q_0 = \{0, 1\} \in S$, we construct a CH sequence u using the following procedure.

   (a) We make k channel assignments for the first frame in u using the following equation:

   $$u_i = \begin{cases} h_0, & \text{if } i \in q_0, \\ h, & \text{if } i \notin q_0. \end{cases}$$

   where h is a randomly selected channel from $\{0, \ldots, N - 1\}$. The (timeslot, channel) assignments for the first frame is obtained using a channel in R, $h_0 = 0$—i.e., $u_0 = 0, u_1 = 0, u_2 = h$.

   (b) Repeat the above procedure to make (timeslot, channel) assignments for each of the other frames using the remaining channels in R (i.e., channels 1 and 2 in this example). The resulting CH sequence is $u = \{0 + 0 \cdot x + h \cdot x^2 + 1 \cdot x^3 + 1 \cdot x^4 + h \cdot x^5 + 2 \cdot x^6 + 2 \cdot x^7 + h \cdot x^8\}$;

4. Repeat Step (3) for each of the other quorums in S (i.e., $q_1 = \{0, 2\}$ and $q_2 = \{1, 2\}$) to construct two other sequences, v and w. The three CH sequences—u, v, w—are the elements of the set Q, which contains $|S| = 3$ CH sequences.

The sequences in Q are illustrated in Fig. 3.2. Algorithm 1 constructs each sequence in Q by making k (timeslot, channel) assignments for each of the m rendezvous channels. One quorum in S is needed to generate each CH sequence in Q. Thus, $|Q| = |S|$.

Note that $\forall u, v \in Q$, there are two corresponding quorums $p, q \in S$ used for constructing u and v, respectively. Because of the intersection property of S, u and v overlap in exactly m distinct channels—viz, the channels $h_d \in R, d \in [0, m - 1]$ (see lines 3 and 6 of Algorithm 1). Also note that all of the sequences in Q have the same period, viz, $T = m \cdot k$ slots. Therefore, Q is a CH system that satisfies the requirements of Problem 1. We refer to the CH system constructed using Algorithm 1 as a *quorum-based channel hopping (QCH)* system.

In the next few sections, we will show how to minimize TTR between any two CH sequences. We will also describe a scheme that addresses the rendezvous convergence problem of CH systems by selecting a particular type of quorum system used in Algorithm 1.

Algorithm 1 QCH System Construction Algorithm

Input: N, k, $R = \{h_0, h_1, \ldots, h_{m-1}\}$, $U = \mathbf{Z}_k$, and a quorum system S under U.
Output: Q.

1: $Q = \emptyset$.
2: **for** $j = 0$ to $(|S| - 1)$ **do**
3:     **for** $d = 0$ to $(m - 1)$ **do**
4:        **for** $i = 0$ to $(k - 1)$ **do**
5:           **if** $i \in q_j$ **then**
6:             $u_{(i+d \cdot m)} = h_d$.
7:           **end if**
8:           **if** $i \notin q_j$ **then**
9:             $u_{(i+d \cdot m)} = h$, randomly chosen from $\{0, \ldots, N - 1\}$.
10:           **end if**
11:        **end for**
12:     **end for**
13:     $Q = Q \cup u$.
14: **end for**

3.4.1 Metrics for Evaluating Channel Hopping Systems

We introduce two metrics—*maximum time-to-rendezvous* (MTTR) and *load*—that are used to evaluate CH systems. Note that both metrics can be used to evaluate all CH systems, not just quorum-based CH systems.

3.4.1.1 Maximum Time-to-Rendezvous

The first metric we introduce is the *maximum time-to-rendezvous* (MTTR) for a CH system, which is defined as the maximum time for any pair of sequences in a CH system to rendezvous. Let $M(H)$ denote the MTTR of a CH system H. It is obvious that the MTTR value impacts the medium access delay of MAC protocols that utilize channel hopping since the exchange of control information is not possible without rendezvous. Networks that require stringent delay requirements will require a CH system with a small MTTR value. For example, in a mobile CR network, neighboring nodes have to exchange time-sensitive control information frequently—information such as spectrum sensing related control information, node location, link connectivity, etc.

In a QCH system Q constructed using Algorithm 1, the MTTR value is equal to the frame length k. In the QCH system Q given in Fig. 3.2, the MTTR is $M(Q) = 3$. In Sect. 3.5.1, we describe a methodology for designing a QCH system that is optimal in terms of minimizing the MTTR.

3.4.1.2 Load of Channel Hopping Systems

Load of Quorum Systems. Here, we provide some definitions regarding the load of quorum systems. We will utilize these definitions to describe one of the optimal QCH systems introduced in Sect. 3.5. In the context of quorum systems, a *strategy* is a rule giving each quorum an access frequency so that the frequencies sum up to one. In other words, a strategy gives the frequency of picking each quorum. A strategy induces a *load* on each element, which represents the fraction of the time the element is used. Specifically, an element's load is the summation of the frequencies of all quorums that the element belongs to. Below, we provide more precise definitions.

Definition 4 Let a quorum system $S = \{q_0, q_1, \ldots, q_{\kappa-1}\}$ be given over a universal set U. Then $W \in [0, 1]^{\kappa}$ is a strategy for S if it is a probability distribution over the quorums $q_j \in S$, i.e., $\sum_{j=0}^{\kappa-1} W_j = 1$.

The *system load*, $\mathcal{L}(S)$, is the minimal load on the busiest element, minimizing over the strategies. A more precise definition follows.

Definition 5 Let a strategy W be given for a quorum system $S = \{q_0, q_1, \ldots, q_{\kappa-1}\}$ over a universal set U. For an element $i \in U$, the load induced by W on i is

$$l_W(i) = \sum_{q_j \in S : i \in q_j} W_j.$$

The *load* induced by a strategy W on a quorum system S is

$$\mathcal{L}_W(S) = \max_{i \in U} l_W(i).$$

The *system load* on a quorum system S is

$$\mathcal{L}(S) = \min_{W} \{\mathcal{L}_W(S)\},$$

where the minimum is taken as the system load over all strategies W.

In CH protocols, spreading out the rendezvous in time and frequency (i.e., channels) is important in order to take full advantage of the frequency diversity of multi-channel medium access. If a large proportion of neighboring nodes rendezvous on the same channel, then channel congestion can occur and lead to a control channel bottleneck problem—we use the term *rendezvous convergence* to refer to such a problem. Some CH protocols (e.g., SSCH [6]) implement "customized" mechanisms to prevent rendezvous convergence. Ideally, a CH protocol should spread out the rendezvous over all channels evenly.

One advantage of using the proposed approach to devise CH schemes is that it can formally characterize the rendezvous convergence problem using the measure of *load* which is used in the study of quorum systems. In quorum systems, a *strategy* is a probabilistic rule that gives the frequency of accessing each quorum so that the frequencies sum up to one. Since a CH system is in essence a quorum system, we can use the definition of load given above to create an analogous definition for the load of a CH system. Let W_0 denote the following strategy: each radio randomly selects a sequence from a CH system with equal probability. Given a CH system, H, the load of H induced by W_0, $\mathcal{L}_{W_0}(H)$, is the load of the busiest element induced by W_0; the busiest element, in this context, refers to the (timeslot, channel)-pair included in the largest number of CH sequences. We define the *load of a CH system* as the load of the CH system induced by the particular strategy W_0.

In the QCH system Q shown in Fig. 3.2, the load is $\mathcal{L}_{W_0}(Q) = 2/3$. In Sect. 3.5, we will discuss a QCH design that is optimal in the sense that it minimizes the load of the QCH system.

3.5 Optimal Synchronous Quorum-Based Channel Hopping Systems

In this section, we describe two optimal QCH systems, both of which require global clock synchronization.

3.5.1 Minimizing the Maximum Time-to-Rendezvous

Minimizing the MTTR of a QCH system is equivalent to minimizing its frame length k. To design an optimal QCH system that minimizes the MTTR, we first need to solve the following problem:

Problem 2 Given a QCH system Q,

$$\begin{aligned} minimize \quad & k, \\ subject\ to \quad & \mathcal{L}_{W_0}(Q) < 1. \end{aligned}$$

The constraint $\mathcal{L}_{W_0}(Q) < 1$ equates to $\bigcap_{u \in Q} u = \emptyset$, which is needed to avoid the scenario in which the load of the QCH system is equal to one (i.e., there is at least one (timeslot, channel)-pair that is included in all of the sequences in Q).

The lower bound for k. To solve Problem 2, we find the lower bound for k when the load of the QCH system is less than one in the following theorem.

Theorem 1 *Given a QCH system Q, a necessary condition for $\mathcal{L}_{W_0}(Q) < 1$ is* $k \geq 3$.

Proof We prove this theorem by contradiction. Let $k \leq 2$ and suppose we have a QCH system Q, where $T = k \cdot m$ and $\mathcal{L}_{W_0}(Q) < 1$.

If $k = 1$, then $T = m$. Since

$$C(u, v) \geq m \geq 1, \forall u, v \in Q,$$

all sequences in Q must be identical. In this case, the load of Q is 1, which contradicts the constraint $\mathcal{L}_{W_0}(Q) < 1$.

If $k = 2$, the universal set is $U = \{0, 1\}$ according to Algorithm 1. Any quorum system S over $U = \{0, 1\}$ has a system load of one, i.e., $\mathcal{L}(S) = 1$. Since the QCH system Q is constructed using S,

$$\mathcal{L}_{W_0}(Q) \geq \mathcal{L}(S) = 1,$$

and this contradicts the constraint $\mathcal{L}_{W_0}(Q) < 1$.

From Fig. 3.2, we can see that there exists a QCH system in which $k = 3$ and its load is less than one.

From the above arguments, it is clear that $k \geq 3$ is a necessary condition for $\mathcal{L}_{W_0}(Q) < 1$.   □

Construction of the M-QCH system. The QCH system that achieves the lower bound for k (i.e., $k = 3$) is an optimal QCH design in the sense that it minimizes the MTTR while keeping the load less than one. We refer to such a system as an *M-QCH System*, and it can be constructed using Algorithm 1 with a *majority* cyclic quorum system over a universal set $U = \mathbf{Z}_3$. The example QCH system shown in Fig. 3.2 is an M-QCH system. An M-QCH system can support m rendezvous channels ($m \in [1, N]$) and it has the lowest MTTR value (i.e., 3) among all QCH systems. Hence, M-QCH systems are advantageous in establishing control channels with minimal medium access delay.

3.5.2 Minimizing the Load

In this subsection, we study an optimal QCH system, Q, that has the minimal load under the constraint $M(Q) \leq \tau$, for a given value of τ. To construct such a QCH system, we need to solve the following problem:

Problem 3 Given a QCH system Q,

$$\begin{aligned} \text{minimize} \quad & \mathcal{L}_{W_0}(Q), \\ \text{subject to} \quad & M(Q) \leq \tau, \end{aligned}$$

where τ is the maximum allowed MTTR of Q.

The lower bound for load. We solve Problem 3 by finding the lower bound for $\mathcal{L}_{W_0}(Q)$ under the constraint $M(Q) \leq \tau$.

Theorem 2 *Given a QCH system Q where $M(Q) \leq \tau$, the minimum load induced by W_0 on Q is $\frac{1}{\sqrt{\tau}}$, i.e.,*

$$\mathcal{L}_{W_0}(Q) \geq \frac{1}{\sqrt{\tau}}.$$

Proof In a QCH system Q where $M(Q) \leq \tau$, since $M(Q) = k$, we have $k \leq \tau$ (k is the frame length).

According to Algorithm 1, such a QCH system Q is constructed using a quorum system S over $U = \mathbf{Z}_k$. Thus, we have $\mathcal{L}_{W_0}(Q) \geq \mathcal{L}(S)$.

The propositions 4.1 and 4.2 in [79] state that the following relation is true:

$$\mathcal{L}(S) \geq \max \left\{ \frac{1}{\varphi(S)}, \frac{\varphi(S)}{k} \right\},$$

where $\varphi(S)$ is the size of the smallest quorum in the quorum system S. Using the inequality of arithmetic and geometric means, it can be shown that

$$\mathcal{L}(S) \geq \frac{1}{\sqrt{k}}.$$

Since $k \leq \tau$, we have $\mathcal{L}_{W_0}(Q) \geq \mathcal{L}(S) \geq \frac{1}{\sqrt{k}} \geq \frac{1}{\sqrt{\tau}}$. $\qquad\square$

Construction of the L-QCH system. Using Algorithm 1, we can construct an optimal QCH system that minimizes the value of $\mathcal{L}_{W_0}(Q)$ by using a *minimal* cyclic quorum system over a universal set $U = \mathbf{Z}_\tau$. We refer to such an optimal QCH system as an *L-QCH System*. An L-QCH system includes τ unique CH sequences. The L-QCH system Q shown in Fig. 3.3 is constructed using a minimal cyclic quorum system S over $U = \mathbf{Z}_7$, which has a degree of overlapping of $m = N$ and $\tau = 7$. The load $\mathcal{L}_{W_0}(Q) = \frac{3}{7} \approx \frac{1}{\sqrt{\tau}}$.

A comparison of all the synchronous CH schemes discussed in this chapter are summarized in Table 3.1.

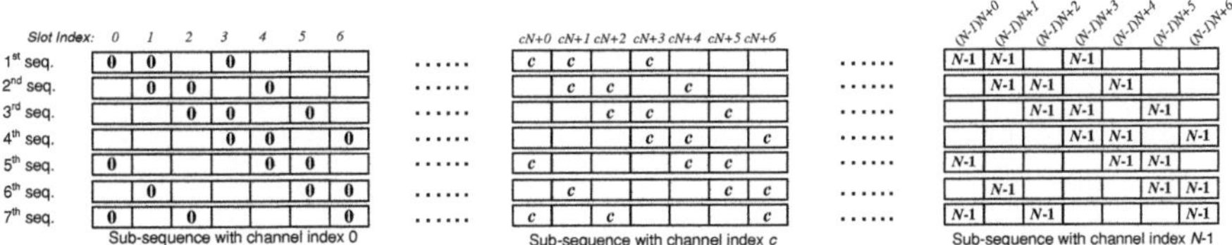

Fig. 3.3 An L-QCH system, Q, with $m = N$ rendezvous channels and $\tau = 7$. To construct Q, we use a minimal cyclic quorum system $S = \{\{0, 1, 3\}, \{1, 2, 4\}, \{2, 3, 5\}, \{3, 4, 6\}, \{4, 5, 0\}, \{5, 6, 1\}, \{6, 0, 2\}\}$ under $U = \mathbf{Z}_7$. The channel indexes for "blank" slots are randomly selected from $[0, N - 1]$. Any two sequences in this system overlap on all N channels

Table 3.1 A comparison of synchronous CH schemes

	Degree of overlapping	MTTR	Load
SSCH	1	$N + 1$	$\frac{1}{N-1}$
M-QCH	$[1, N]$	3	$\frac{2}{3}$
L-QCH	$[1, N]$	k	$\frac{1}{\sqrt{\tau}}$

3.5.3 Performance Evaluation

We simulate the QCH scheme in network simulator [104] and use three MAC-layer reference protocols for comparison: IEEE 802.11b, SSCH, and the SeqR protocol. The data rate is 11 Mbps by default. Note that IEEE 802.11b and SSCH were not designed for use in CR networks. However, they serve as good benchmarks in evaluating the performance of QCH. Furthermore, the design criteria of CH schemes for conventional multi-channel networks are almost identical to those of CH schemes for CR networks, and SSCH is one of the most well-known schemes of the former type. In the simulations, secondary nodes can opportunistically access three channels (i.e., $N = 3$). The channel switching delay is chosen as 80 μs, which is well supported by existing technology [30]. The duration of a time slot is 200 ms. Every node uses Ad hoc On-Demand Distance Vector Routing (AODV) [80] as the routing protocol. At the transport layer, UDP is used in the simulations by default. The traffic generator uses Constant Bit Rate (CBR) flows with a flow rate of 11 Mbps and a packet size of 512 bytes. The transmission range of every node is 250 m. We simulated two networks: a static network with ten single-hop flows in a $100\,\text{m} \times 100\,\text{m}$ square area and a random network with five multi-hop flows in a $1000\,\text{m} \times 1000\,\text{m}$ square area. In the simulations, we study the time-to-rendezvous (TTR) value and the throughput in each CH protocol under varying conditions, including random incumbent traffic, node mobility, and multi-flow complex networks.

The following protocols were simulated.

- SSCH: Each node randomly chooses one (channel, seed)-pair to construct its CH sequence. For example, if a node selects the pair $(0, 1)$, then its CH sequence has a period of $(N + 1)$ slots, and the channel indexes in its sequence are $\{0, 1, 2, 1\}$. The last slot of a period is the parity slot, and the channel index of this slot is equal to the value of the node's seed.
- SeqR: This is the protocol proposed in [23]; it was briefly described in Sect. 3.1.
- QCH: We simulate two synchronous QCH systems (M-QCH, and L-QCH).

We assume that every node randomly picks one sequence from a CH system and performs channel hopping in accordance with the sequence. Once the sending-receiving node pair rendezvous on a channel, the pair performs common hopping to exchange data packets. The sender follows the receiver's sequence.

Incumbent traffic generation. In the simulations, we generated incumbent traffic as follows. In every time slot, the incumbent transmitter decides whether to transmit or not by flipping a coin. If the incumbent transmitter decides to transmit, it randomly

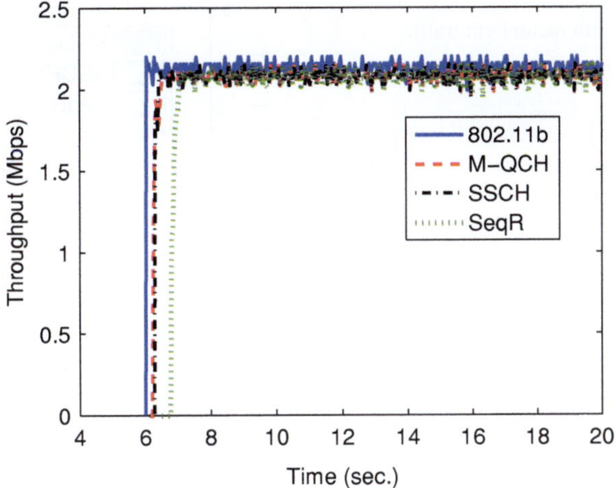

Fig. 3.4 Effect of time-to-rendezvous and channel switching

selects one or two channels and transmits packets in the current time slot. All of the secondary nodes are within the transmission range of the incumbent transmitter. A single incumbent transmitter was simulated. A channel is tagged as "unavailable" while incumbent traffic is present on it. All secondary nodes should refrain from transmitting on unavailable channels during the period of incumbent transmission. Note that all nodes that perform channel hopping are secondary nodes.

3.5.3.1 Impact of Time-to-Rendezvous

We first simulated a single-hop flow to show the effect of TTR on channel access delay and the effect of channel switching overhead on throughput. The results are shown in Fig. 3.4. We can see that the starting times of the traffic delivery for the simulated protocols are different, which coincides with the discrepancy of the protocols' channel access delays due to the variation in TTR values. Note that the throughput of each CH protocol is lower than that of 802.11b, which we can attribute to the channel switching overhead.

Next, we simulated a network with ten single-hop *disjoint* flows in a $100\,\mathrm{m} \times 100\,\mathrm{m}$ square area. Two flows are considered disjoint if they do not share either endpoint. The average TTR for three CH schemes (when there is no incumbent traffic) is shown using the leftmost group of bars in Fig. 3.5. As can be seen, M-QCH has the lowest average TTR compared to SSCH and SeqR—this is expected since M-QCH has the lowest MTTR value among the three CH protocols.

3.5.3.2 Impact of Degree of Overlapping

As expected, our simulation results indicate that a CH scheme's degree of overlapping has a clear impact on its TTR value when the incumbent transmitter is active. The average TTR for three CH schemes in the presence of incumbent traffic is shown using

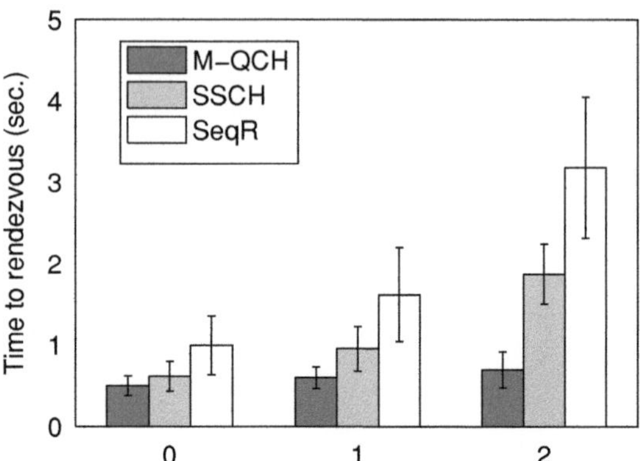

Fig. 3.5 Time-to-rendezvous with incumbent traffic

the center and rightmost groups of bars in Fig. 3.5. M-QCH has a clear advantage over SSCH and SeqR in terms of TTR, because M-QCH's degree of overlapping is greater than that of the other two schemes in this simulation. This advantage becomes more evident in the presence of incumbent traffic since a pair of nodes using M-QCH can rendezvous on other channels if the current rendezvous channel is occupied by incumbent signals. In contrast, a pair of nodes using either SSCH or SeqR can rendezvous only on one channel (here, we are referring to the initial rendezvous). This implies that the nodes may not be able to achieve the initial rendezvous until the incumbent vacates the rendezvous channel. Note that in SSCH, initial rendezvous is needed to exchange data, such as each other's sequence, that is required to rendezvous in multiple channels.

Next, we set up ten *non-disjoint* flows in a $100\,\mathrm{m} \times 100\,\mathrm{m}$ square area, where every node serves as both a transmitter and a receiver in multiple flows. In other words, this scenario includes multiple simultaneous flows with common endpoints. We assume temporary common hopping, i.e., each transmitter node has to change its hopping sequence and follow the receiver's sequence after a rendezvous has occurred to bootstrap communications. If the receiver node also acts as a transmitter in another flow, it must also follow the sequence of its intended receiver after a rendezvous. Thus, some nodes in this network have to switch between CH sequences continuously, and this scenario puts stress on a CH protocol's ability to establish links. Temporary common hopping prescribes that a transmitter should return back to its original sequence once the transmission has finished—this avoids the global synchronization of the CH sequences over the entire network (i.e., a scenario in which every node uses the same sequence). We compare the per-flow throughput of M-QCH, SSCH, and SeqR in Fig. 3.6. When an incumbent signal is detected, M-QCH replaces the incumbent-occupied channel in its sequence with any incumbent-free channel. The results in the figure show that M-QCH outperforms the other two protocols, since M-QCH is faster in re-establishing links—this, of course, is due to the fact that M-QCH enables rendezvous in a greater number of distinct channels per sequence period.

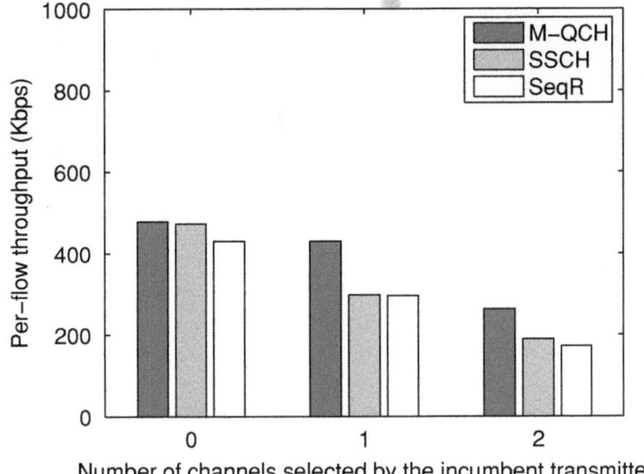

Fig. 3.6 Effect of load on throughput. **a** Per-flow throughput. **b** System throughput

3.5.3.3 Impact of the Load

In this set of simulation experiments, we investigated the effect of load (as defined in Sect. 3.4.1.2) on network performance. In general, the load of a CH system determines the number of concurrent co-channel transmissions in each time slot. If the load is low, the number of concurrent co-channel transmissions in each slot is small, which means that the traffic is more evenly distributed among different channels in every time slot. In general, a more even distribution of traffic among channels implies higher network throughput. In the simulations, we varied the number of disjoint flows in a $100\,\text{m} \times 100\,\text{m}$ square area, and the results are shown in Fig. 3.7. In the figures, we can see that L-QCH and SeqR outperform the other schemes since the two schemes have the lowest load (viz $\frac{1}{3}$) compared to the other schemes (M-QCH has a load of $\frac{2}{3}$ and SSCH incurs a load of $\frac{1}{2}$ as indicated in Table 3.1). It is interesting to note that M-QCH's performance is inferior to that of the other schemes when the number of flows is small (because M-QCH has the highest load); however, when the number of flows is large, the system throughput for SSCH is lower than that of the other schemes. This phenomenon can be attributed to the limiting effect of the parity slot prescribed by SSCH when the network is close to saturation: nodes using SSCH can only utilize $(N-1)$ channels specified by $(N-1)$ seed values in the parity slot. In contrast, nodes using the other CH protocols can fully utilize all N channels in any time slot. From Fig. 3.7, we can conclude that CH schemes that have lower load values are generally advantageous in terms of being able to support higher throughput; when the network is nearly saturated, the system throughput is closely related to the number of channels that can be fully utilized by each CH scheme.

3.6 Asynchronous Channel Hopping System

In Sect. 3.4, by utilizing the intersection property of quorum systems, a set of CH sequences can be generated such that any pair selected from the set is guaranteed to rendezvous when the boundaries of the two selected CH sequences are synchronized.

Fig. 3.7 Throughput of non-disjoint flows with incumbent traffic

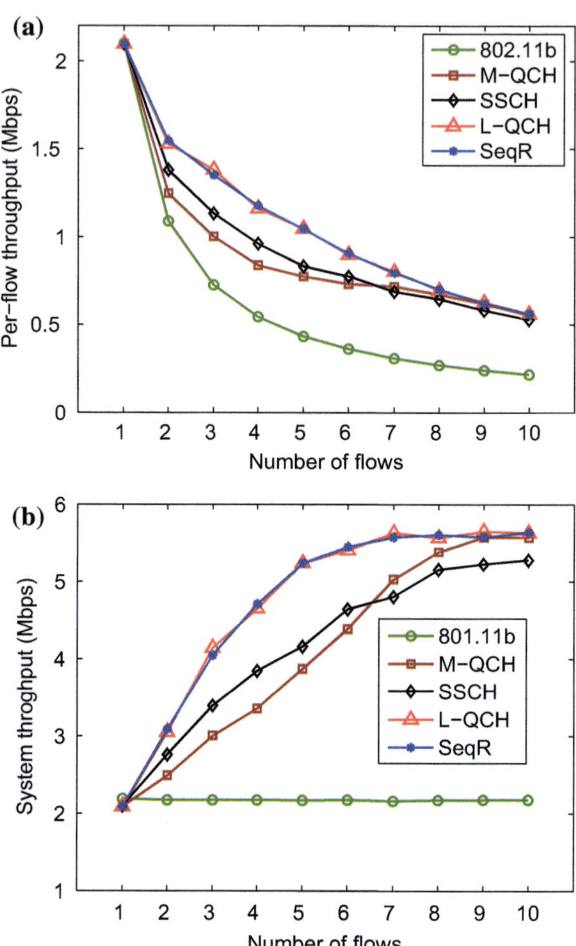

However, merely relying on the intersection property of quorum systems is *not* sufficient to guarantee rendezvous when the two CH sequences are *asynchronous* (i.e., the boundaries of the two CH sequences are misaligned).

The asynchronous channel hopping based rendezvous schemes can be classified into two categories.

- Asymmetric design: every node has a priori knowledge of its role as either a sender or a receiver (e.g., Bluetooth pairing). The sender and receiver nodes use *different* methods to generate their CH sequences for achieving rendezvous.
- Symmetric design: every node behaves identically (i.e., a node may be either a sender or a receiver) and follows the *same* method for generating its CH sequence.

Since the symmetric design is independent of the assumption that every node has a priori knowledge of its role as either a sender or a receiver, it has a wider variety of applications than the asymmetric design. Note that a CH scheme can be either a symmetric or an asymmetric design depending on whether the sender and receiver employ different CH sequence generation algorithms or hop at different rates. For example, if every node has the same hopping rate [91, 100] or the same sequence-generation algorithm [6, 11, 23, 93], then the CH scheme is a symmetric design. In contrast, if the sender and receiver have different hopping rates [97] or employ different sequence-generation algorithms [10], then the CH scheme is an asymmetric design.

We first introduce the concepts of *cyclic rotation* and the *rotation closure property* in the context of quorum systems. Given a non-negative integer k and a quorum q in a quorum system Q under the universal set $U = \{0, \ldots, n-1\}$, we use $rotate(q, k)$ to denote a *cyclic rotation* to quorum q:

$$rotate(q, k) = \{(i + k) \bmod n \,|\, i \in q\}.$$

For example, suppose we have a quorum system $Q = \{\{0, 1\}, \{1, 2\}\}$ under $U = \mathbf{Z}_3$. For quorum $q = \{0, 1\}$ and $k = 2$, $rotate(q, k) = \{2, 0\}$. Now let us define the *rotation closure property* of quorum systems.

Definition 6 A quorum system Q over $U = \mathbf{Z}_n$ is said to have the rotation closure property if the following holds:

$$\forall p, q \in Q, \forall k \in [0, n-1], rotate(p, k) \cap q \neq \emptyset.$$

For the sake of simplicity, we represent a CH sequence u of period T as a set of channel indexes:

$$u = \{u_0, u_1, \ldots, u_i, \ldots, u_{T-1}\},$$

where $u_i \in [0, N-1]$ represents the channel index of sequence u in the ith timeslot of a CH period.

We extended the concept of cyclic rotation so that it is applicable to CH sequences. Given a CH sequence u, we use $rotate(u, k)$ to denote a *cyclic rotation* of CH sequence u by k timeslots, i.e.,

$$rotate(u, k) = \{v_0, \ldots, v_j, \ldots, v_{T-1}\},$$

where $v_j = u_{(j+k) \bmod T}$, $j \in [0, T-1]$ and k is a non-negative integer. For example, given $u = \{0, 1, 2\}$ and $T = 3$, $rotate(u, 2) = \{2, 0, 1\}$.

A channel hopping (CH) system of period T is defined as a set of CH sequences of period T. Here, we formally define an *asynchronous channel hopping* (ACH) system below.

Definition 7 An asynchronous channel hopping (ACH) system of period T is a CH system of period T, which satisfies the *rotation closure property*: it consists of CH sequences such that any two distinct CH sequences u and v satisfy the following inequality:

$$\forall k, l \in [0, T-1], C(rotate(u, k), rotate(v, l)) \geq m,$$

where the positive integer m is *the degree of overlapping* of the ACH system.

If the sequences of a sender and a receiver belong to an ACH system, the two nodes can rendezvous with each other on at least m distinct rendezvous channels even if their clocks are asynchronous.

The following theorem states that a CH system with the rotation closure property ensures rendezvous even when the slot boundaries or CH sequence boundaries are not aligned.

Theorem 3 *If a CH system H of period T with a degree of overlapping m satisfies the rotation closure property, any pair of CH sequences in H must overlap by at least $m/2$ timeslots for every T consecutive timeslots even when the timeslot boundaries are misaligned by an arbitrary amount.*

Proof Suppose that a CH system H satisfies the rotation closure property and two nodes, A and B, each picks a CH sequence from H randomly—viz, u and v, respectively. For the sake of our discussions, suppose the length of a timeslot is 1. We consider two cases.

(1) When slot boundaries are aligned: Without loss of generality, let us suppose node A's clock is i slots ahead of node B's clock. With respect to node B's clock, node A's sequence u is equivalent to $rotate(u, i)$. Since H has the rotation closure property, $C(rotate(u, i), v) \geq m$. Hence, the two sequences must have at least m rendezvous channels between them (i.e., overlap by at least m timeslots). It is obvious that the same result is obtained when we assume that A's clock is i slots behind B's clock.

(2) When slot boundaries are misaligned: Suppose node A's clock is ahead of node B's clock by an arbitrary amount of time, say $i + \delta$, where $i \in \mathbf{Z}_T, 0 < \delta < 1$.

(a) If $\delta \leq 1/2$, let us shift left node B's sequence by δ and designate this sequence as v'.[2] It is obvious that the slot boundaries of u and v' are aligned and the former is ahead of the latter by i slots in terms of their respective nodes' clocks. Since H has the rotation closure property, $C(rotate(u, i), v') \geq m$, and thus u must overlap with v by $m(1-\delta)$ timeslots. This means that the two sequences overlap with each other by at least $m/2$ timeslots for every T consecutive timeslots.

(b) If $\delta > 1/2$, let us shift right node B's sequence by $1 - \delta$ and designate this sequence as v'. It is obvious that the slot boundaries of u and v' are aligned and the former is ahead of the latter by $i + 1$ slots in terms of their respective nodes' clocks. Since H has the rotation closure property, $C(rotate(u, i+1), v') \geq m$, and thus u must overlap with v by $m\delta$ timeslots. This means that the two sequences overlap with each other by at least $m/2$ timeslots for every T consecutive timeslots.   □

---

[2] Shifting a node's CH sequence *left/right* by δ is equivalent to *advancing/retreating* the node's clock by δ.

From Theorem 3, we can conclude that any two nodes that select CH sequences from a system with the rotation closure property can rendezvous with each other during the overlap of their sequences even if they are asynchronous (i.e., slot boundaries are not aligned). If multiple pairs of nodes happen to rendezvous at the same slot on the same channel, they can follow a channel contention procedure (e.g., 802.11 RTS/CTS protocol) to carry out the pairwise rendezvous.

The ACH System Design Problem. In the OSS paradigm, the appearance of primary user signals can hamper the secondary user nodes' rendezvous process, since primary users can occupy any given channel and have absolute priority over all channels. To minimize the chance of rendezvous failure due to those factors, we would ideally want an ACH system with the following property: in an OSS environment with N available channels, an ACH system of period T guarantees N distinct rendezvous channels between any two CH sequences for every T consecutive timeslots, i.e., its degree of overlapping is $m = N$.

Our research findings indicate that only certain types of quorum systems satisfy the *rotation closure property* and that only such quorum systems can guarantee pairwise rendezvous between two asynchronous sequences. An example is given below.

- Suppose we have a quorum system Q, under the universal set $U = \mathbf{Z}_n$, that satisfies the rotation closure property. Two quorums $p, q \in Q$ must have an intersection even if an arbitrary amount of cyclic rotation is applied to either one of them.
- Suppose we have an ACH system that includes two sequences of period $T = n$, i.e., u and v. Given a channel index $h \in [0, N-1]$, we assign the channel index h to u and v in the following way: let $u_i = h$ if $i \in p$ and let $v_i = h$ if $i \in q$, where $i \in [0, T-1]$. It is obvious that the rotation closure property of Q guarantees that the two sequences rendezvous on channel h even if an arbitrary amount of cyclic rotation is applied to either of the sequences.

3.6.1 The Asynchronous Channel Hopping System with a Degree of Overlapping m = 2

In this section, we describe an *asynchronous* CH system that does *not* require global clock synchronization. The objective is to devise a CH system, H, that enables any pair of CH sequences to overlap by at least half of a timeslot for every sequence period (i.e., for every T consecutive timeslots) even under the assumption that slot boundaries are misaligned by an arbitrary amount.

Next, we describe an algorithm, Algorithm 2, for constructing an ACH system that uses two different types of cyclic quorum systems. Note that the constructed ACH system has a degree of overlapping $m = 2$, and this it is referred to as an asynchronous QCH (AQCH) system.

Construction of AQCH Systems. Algorithm 2 uses two types of cyclic quorums systems to construct an AQCH system that guarantees at least two rendezvous channels between any two sequences (i.e., $C(u, v) \geq 2, \forall u, v \in H$). This AQCH system

---

Algorithm 2 AQCH System Construction Algorithm

Input: $N, k, R = \{h_0, h_1\}, U = \mathbf{Z}_k$, and two cyclic quorum systems S and S' over U.
Output: Q.
1: $Q = \emptyset$.
2: **for** $j = 0$ to $(|S| - 1)$ **do**
3:     **for** $i = 0$ to $(k - 1)$ **do**
4:         **if** $i \in B_j$ **then**
5:             $u_i = h_0$.
6:         **else if** $i \in B'_j$ **then**
7:             $u_i = h_1$.
8:         **else**
9:             $u_i = h$ randomly chosen from $\{0, \ldots, N - 1\} \setminus \{h_0, h_1\}$.
10:         **end if**
11:     **end for**
12:     $Q = Q \cup u$.
13: **end for**

---

is composed of CH sequences that have only one frame per sequence period (i.e., $T = k$). In each constructed CH sequence, a subsequence constructed by a minimal cyclic quorum S is interleaved with a subsequence constructed by a majority cyclic quorum system S'. Refer to [68] for methods to construct minimal cyclic quorums and majority cyclic quorums. A pseudo-code of Algorithm 2 is given above and its description is given below.

1. First construct a universal set $U = \mathbf{Z}_k$;
2. Find a minimal (k, κ)-difference set $D = \{a_1, a_2, \ldots, a_\kappa\}$ such that $\kappa < \frac{k}{2}$ and construct a minimal cyclic quorum system based on D, such as $S = \{B_i | B_i = \{a_1 + i, a_2 + i, \ldots, a_\kappa + i\} \mod k, i \in [0, k - 1]\}$;
3. Construct a relaxed cyclic $(k, \lceil\frac{k+1}{2}\rceil)$-difference set $D' = \{a'_1, a'_2, \ldots, a'_{\lceil\frac{k+1}{2}\rceil}\}$ such that $D' \cap D = \emptyset$. Then, construct a majority cyclic quorum system based on D', such as $S' = \{B'_i | B'_i = \{a'_1 + i, a'_2 + i, \ldots, a'_{\lceil\frac{k+1}{2}\rceil} + i\} \mod k, i \in [0, k - 1]\}$.
   Note that $|S| = |S'| = k$, and $|D'| = \lceil\frac{k+1}{2}\rceil$;
4. Use the minimal cyclic quorum system S for assigning the first rendezvous channel h_0 to appropriate slots in a CH sequence u (see lines 4 and 5 in the algorithm);
5. Use the majority cyclic quorum system S' to assign the second rendezvous channel h_1 to appropriate slots in a CH sequence u (see lines 6 and 7 in the algorithm).
6. The remaining slots in u are assigned a channel index randomly chosen from $\{0, \ldots, N - 1\} \setminus \{h_0, h_1\}$ (see line 9 in the algorithm).

An example ACH system is shown in Fig. 3.8. One can readily show that an ACH system constructed using Algorithm 2 satisfies the rotation closure property and that the TTR value between any two sequences in an ACH system is bounded by the length of its sequence period $T = k$.

Fig. 3.8 An ACH system Q with $m = 2$ and $T = k = 9$. The universal set, U, is $\mathbf{Z}_9$, S is constructed using $D = \{0, 1, 2, 4\}$, and S' is constructed using $D' = \{3, 5, 6, 7, 8\}$. Note that $D' \cap D = \emptyset$. The numbers inside the slots represent the channel index values

Slot Index:	0	1	2	3	4	5	6	7	8
	0	0	0	1	0	1	1	1	1
	1	0	0	0	1	0	1	1	1
	1	1	0	0	0	1	0	1	1
	1	1	1	0	0	0	1	0	1
	1	1	1	1	0	0	0	1	0
	0	1	1	1	1	0	0	0	1
	1	0	1	1	1	1	0	0	0
	0	1	0	1	1	1	1	0	0
	0	0	1	0	1	1	1	1	0

Given that $D' \cap D = \emptyset$, k must be no less than $|D| + |D'|$ so that a CH sequence can accommodate two subsequences constructed using S and S'. In [68], Luk and Wong conducted an exhaustive search to find the minimal difference sets under $\mathbf{Z}_k$ for $k = 4, \ldots, 111$.

In our description of ACH systems given above, we used relaxed cyclic difference sets D and D' for generating cyclic quorum systems that facilitate the construction of a CH system with the rotation closure property. The specific choices of D and D' and the resulting cyclic quorum systems have no significance—i.e., the quorum systems that we have chosen are merely our design choices for constructing an ACH system; it is likely that there are other quorum systems that can be used to construct ACH systems of similar or different structure.

With two *disjoint* cyclic difference sets under $\mathbf{Z}_T$, one can readily construct two cyclic quorum systems. These quorum systems can be used to generate CH sequences of period T that have *two* pairwise rendezvous channels, even in a time asynchronous environment. To construct an ACH system of period T that enables N rendezvous channels between CH sequences, we have to find N disjoint cyclic difference sets under $\mathbf{Z}_T$. However, identifying N disjoint cyclic difference sets under $\mathbf{Z}_T$ is a difficult problem in general. In fact, in many cases, N disjoint cyclic difference sets under $\mathbf{Z}_T$ may not exist for some given values of N and T. For example, the number of disjoint cyclic difference sets under $\mathbf{Z}_{12}$ when $N = 3$ and $T = 12$ is less than three. Thus, we will show how to construct an ACH with the maximum degree of overlapping using the second category of quorum systems, i.e., the array-based quorum system.

3.6.2 The Asynchronous Channel Hopping System with Maximum Degree of Overlapping

3.6.2.1 Array-Based Quorum Systems

In [56], the authors identified a few types of quorum systems that have the rotation closure property. We classify these quorum systems into two categories based on their construction methods. The first category of quorum systems is the cyclic quorum system that can be constructed using cyclic difference sets, which we have introduced in Sect. 3.2. Then, we introduce the second category of quorum systems.

This category of quorum systems can be constructed using an $r \times l$ array, $A[\cdot][\cdot]$. The universal set U is $\{0, \ldots, r \cdot l - 1\}$, and there is a one-to-one mapping between the element $(i \cdot l + j)$ where $i \in [0, r - 1]$, $j \in [0, l - 1]$ in the universal set and the array element $A[i][j]$.

- In the *square* quorum system [68], a quorum is constructed by picking a full column and a full row of elements from the array.
- In the *torus* quorum system [56], a quorum is constructed by picking a full column $c \in [0, l - 1]$ plus $\lfloor l/2 \rfloor + 1$ elements, each of which is located in distinct columns $(c + i) \bmod l, i = 0, \ldots, \lfloor l/2 \rfloor$.
- We propose a new type of array-based quorum system that extends the torus quorum system. In this quorum system, a quorum is constructed by picking a full column of elements, $c \in [0, l - 1]$, plus l elements, each one located in a different column of the array. Since any quorum in this array-based quorum system is a super set of a torus quorum, the proposed array-based quorum system satisfies the rotation closure property.

As stated above, a quorum in an array-based quorum system must contain a full column of elements and k elements ($k = l$ or $k = \lfloor l/2 \rfloor + 1$) selected from different columns of the array. We call the full column of elements as the *column* of the quorum and the other k elements as the *span* of the quorum. Note that the column and the span of the same quorum have a common element.

An example of the array-based quorum system is illustrated in Fig. 3.9. In the example, given two quorums, p and q, the column of p must have an intersection with the span of q, even if a "cyclic rotation" is applied to either p or q, and vice versa. The cyclic rotation to q will generate another quorum, q', of the same type. These properties hold for any two quorums in an array-based quorum system.

3.6.2.2 Constructing Asynchronous Channel Hopping Systems

In this subsection, we introduce an algorithm that uses an array-based quorum system to construct an ACH system that can be used by secondary nodes to establish links in a distributed manner. Suppose we want to construct an ACH system that enables any pair of CH sequences to rendezvous in N distinct channels, where N is the total number of available channels. The algorithm generates a set of CH sequences of period $T = r \cdot l$, given an $r \times l$ array where $r, l \geq N$. The reasons for utilizing an array-based quorum system are:

- The intersections between the columns and the spans enables the pair of sequences (i.e., sender's and receiver's sequences) to rendezvous on every one of the N channels;

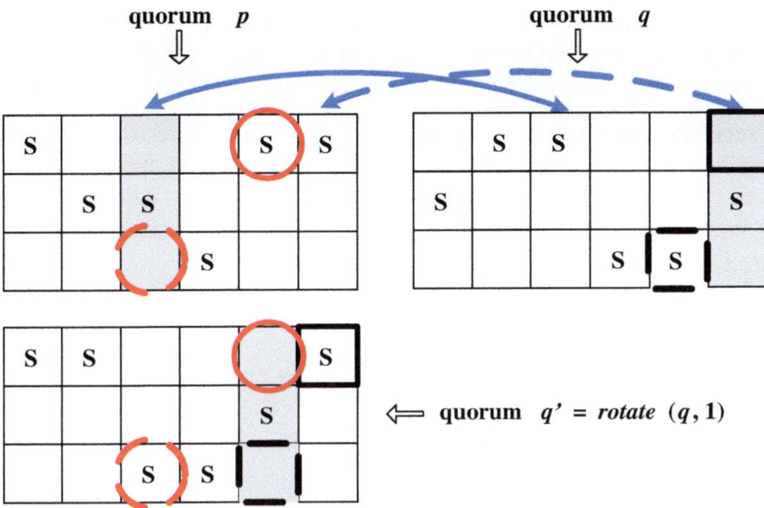

Fig. 3.9 Two quorums, p and q, each with 8 elements, in an array-based quorum system. Both quorums are constructed using a 3×6 array. An array-based quorum includes a column of elements (*squares* colored in *grey*) and a span of elements (*squares* labeled with a 'S'). A span and a column have one element in common which is not counted twice. One quorum's column must have an intersection with a span of another. For instance, in the figure, p's column intersects q's span at the *square* that is located in the first row of the third column. Similarly, p's span intersects q's column at the *square* located in the first row of the sixth column. Note that performing cyclic rotation by one slot to either of the array-based quorums yields a third array-based quorum (e.g., $q' = rotate(q, 1)$), and the intersection property is maintained after the cyclic rotation. The elements in common between p and q' are marked with *circle* markings, and those between q and q' are marked with *square* markings

- The rotation closure property of the array-based quorum system guarantees rendezvous on every one of the N channels even if a cyclic rotation is applied to either of the two sequences.

The sender (or the receiver) generates an array-based quorum system using an $r \times l$ array, assigns channel index values to quorums' columns (or spans), and derives the CH sequence from the channel index assignment. The procedure for constructing the sequences is described below, and the corresponding pseudo code is given in Algorithm 3.

1. The sender constructs an $r \times l$ array, $S[\cdot][\cdot]$, which stores the channel index values assigned to columns' elements.

    (a) The sender randomly assigns N channel indexes to the l columns of $S[\cdot][\cdot]$ such that each channel index is assigned to at least one column. All array elements in the same column are assigned the same channel index.

(b) Let h_i denote the channel index assigned to the i-th column, where $i \in [0, l-1]$. The sender derives its CH sequence, u, in the following way: $u_{i \cdot l + j} = h_j$, where $i \in [0, r-1]$ and $j \in [0, l-1]$.

2. The receiver constructs an $r \times l$ array, $R[\cdot][\cdot]$, which stores the channel index values assigned to spans' elements. On basis of $R[\cdot][\cdot]$, an array-based quorum system can be constructed, from which the receiver chooses r quorums that have r *disjoint* spans. Let s_k denote the span that contains the array element $R[k][0]$ where $k \in [0, r-1]$ (see line 14 in Algorithm 3).

   (a) The receiver randomly assigns N channel indexes to the r spans such that each channel index is assigned to at least one span, and all array elements belonging to the same span are assigned the same channel index.
   (b) Let h'_k denote the channel index assigned to the span s_k. The receiver derives its CH sequence, v, in the following way: $v_{i \cdot l + j} = h'_k$ if $R[i][j] \in s_k$ where $i \in [0, r-1]$ and $j \in [0, l-1]$.

Using Algorithm 3, we can assign the channel indexes to the sender's and receiver's CH sequences. We refer to $S[r][l]$ and $R[r][l]$ with their array elements fully assigned as the *sequence array of the sender* and the *sequence array of the receiver*, respectively. In the above method, channel indexes are assigned to array elements in a column-wise manner by the sender, or a span-wise manner by the receiver. Hence, we call the sender's CH sequence u as a *column-based* sequence, and the receiver's CH sequence v as a *span-based* sequence. Figure 3.10 illustrates an example ACH system built using a 3×5 array when $N = 3$. In the example, the degree of overlapping between the two sequences is N.

Theorem 4 *Algorithm 3 generates an ACH system, $H = \{u, v\}$, such that the pair of CH sequences rendezvous in N distinct channels within a sequence period, even if the two sequences are misaligned by an arbitrary amount.*

Proof For the sake of our discussions, suppose the length of a timeslot is 1. We prove this theorem in two scenarios.

Case 1: Slot boundaries aligned. As previously mentioned, Algorithm 3 generates the sender's and receiver's CH sequences using two $r \times l$ arrays. Without

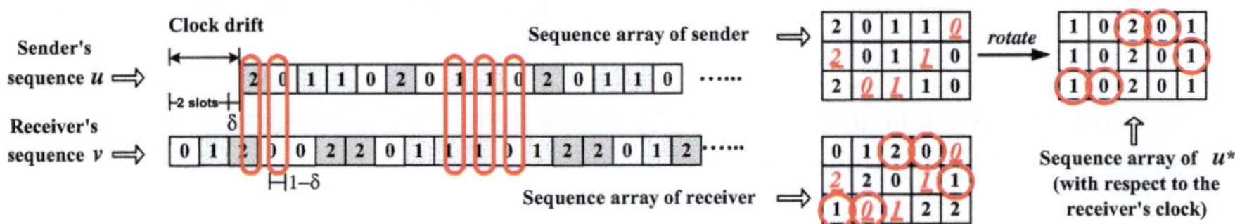

Fig. 3.10 An ACH system H when $N = 3$, $r = 3$, $l = 5$, and the receiver's clock is ahead of the sender's clock by $(2 + \delta)$ timeslots where $\delta \leq 1/2$. The sequence arrays of the sender and the receiver that generated the respective sequences are also shown. It can be easily seen that the two CH sequences have N distinct rendezvous channels despite the clock drift and that the overlap duration on each rendezvous channel is $(1 - \delta)$

Algorithm 3 ACH System Construction Algorithm

Input: N, r, l.

Output: $H = \{u, v\}$.

1: $D = D' = \{0, 1, \ldots, N - 1\}$.
2: $S[r][l] \leftarrow$ sequence array of the sender.
3: $R[r][l] \leftarrow$ sequence array of the receiver.
4: **for** $j = 0$ to $l - 1$ **do**
5:      **if** $D = \emptyset$ **then**
6:          h_j is randomly chosen from $\{0, 1, \ldots, N - 1\}$.
7:      **else**
8:          Randomly select h_j from D, $D \leftarrow D \setminus \{h_j\}$.
9:      **end if**
10:      **for** $i = 0$ to $r - 1$ **do**
11:          $u_{i \cdot l + j} = h_j$.
12:      **end for**
13: **end for**
14: Choose r array-based quorums that have r disjoint spans, $s_0, \ldots, s_k, \ldots, s_{r-1}$, such that $R[k][0] \in s_k$.
15: **for** $i = 0$ to $r - 1$ **do**
16:      **if** $D' = \emptyset$ **then**
17:          h'_i is randomly chosen from $\{0, 1, \ldots, N - 1\}$.
18:      **else**
19:          Randomly select h'_i from D', $D' \leftarrow D' \setminus \{h'_i\}$.
20:      **end if**
21: **end for**
22: **for** $i = 0$ to $r - 1$ **do**
23:      **for** $j = 0$ to $l - 1$ **do**
24:          **for** $k = 0$ to $r - 1$ **do**
25:              **if** $R[i][j] \in s_k$ **then**
26:                  $v_{i \cdot l + j} = h'_k$.
27:              **end if**
28:          **end for**
29:      **end for**
30: **end for**

---

loss of generality, let us suppose the receiver's clock is i slots ahead of the sender's clock, where i is an integer. Suppose that their timeslot boundaries are aligned. With respect to the receiver's clock, sender's sequence u is equivalent to $rotate(u, i)$, and the operation $rotate(u, i)$ yields a new CH sequence u^*. In the sequence array of u^*, all the elements in each column are assigned the same channel index, and thus l columns contain N channel indexes.

If the sequence array of u^* and the sequence array of the receiver (see the example in Fig. 3.10) are superimposed on top of one another, every column of the former overlaps with one of the spans of the latter. Since $r, l \geq N$, one can readily observe that there are N overlaps of N distinct channel indexes. An overlap occurs when the same channel index appears in the same position (i.e., same row and column positions) in the two arrays. Thus, u^* and v have N distinct rendezvous channels within a sequence period.

Case 2: Slot boundaries misaligned. In this case, suppose the receiver's clock is ahead of the sender's clock by an arbitrary amount of time, say $(i + \delta)$, where $0 < \delta < 1$.

- If $\delta > 1/2$, let us shift right the sender's sequence by $1 - \delta$ and designate this sequence as u'. It is obvious that the timeslot boundaries of u' and v are aligned and the latter is ahead of the former by $(i + 1)$ slots in terms of their respective nodes' clocks. In Case 1, we have shown that $C(rotate(u, i), v) = N$ for any integer i. Since $\delta < 1$, u' and v rendezvous in N distinct channels within a sequence period and the duration of overlap on each rendezvous channel is δ.
- If $\delta \leq 1/2$ (the case shown in Fig. 3.10), let us shift left the sender's sequence by δ and designate this sequence as u'. It is obvious that the timeslot boundaries of u' and v are aligned and the latter is ahead of the former by i slots in terms of their respective nodes' clocks. Since $C(rotate(u, i), v) = N$ for any integer i and $\delta < 1$, u' and v rendezvous in N distinct channels within a sequence period and the duration of overlap on each rendezvous channel is $(1 - \delta)$. $\square$

By proving Theorem 4 we have validated that Algorithm 3 generates a pair of CH sequences that can be used by two channel hopping nodes to solve the asynchronous rendezvous problem in the context of opportunistic spectrum sharing. Note that the ability to rendezvous in N distinct channels is a critical feature—it minimizes the chance of rendezvous failure due to appearance of primary user signals.

3.6.3 Optimal Asynchronous Channel Hopping Systems

In this section, we first introduce the metric for evaluating CH schemes that do not require clock synchronization—*minimum rendezvous probability* (MRP). Then, we formulate the optimal ACH design problem for CR networks.

3.6.3.1 Minimum Rendezvous Probability

The *minimum rendezvous probability* (MRP) of an ACH system is defined as the lower bound of the probability that a pair of asynchronous sequences from an ACH system will rendezvous in a given timeslot. In an ACH system H with period T and a degree of overlapping value of m, two sequences are guaranteed to rendezvous at least m times on m distinct rendezvous channels during a sequence period. Thus, we define the MRP for an ACH system H as,

$$P(H) = \frac{m}{T}. \tag{3.1}$$

3.6.3.2 The Optimal Asynchronous Channel Hopping Design Problem

As stated previously, an ACH system that can serve as a solution to the asynchronous rendezvous problem is a system whose degree of overlapping is equal to the total number of available channels, viz N. The MRP of such a system is $\frac{N}{T}$, and maximizing the MRP for a fixed value of N is equivalent to minimizing the sequence period, T. We define an *optimal* ACH system as a system whose degree of overlapping is N and has the maximum possible MRP value. To devise an optimal ACH system, one has to solve the following optimization problem:

Problem 4 Given an ACH system $H = \{u, v\}$ of period T and assuming there are N available channels,

$$\begin{aligned}
&\text{minimize} \quad T, \\
&\text{subject to} \quad \forall k, l \in [0, T-1], \\
&\qquad\qquad C(rotate(u, k), rotate(v, l)) = N.
\end{aligned}$$

The solution to Problem 4 is given by the following theorem.

Theorem 5 *In an ACH system H whose degree of overlapping is N, the period of any CH sequence must be N^2 or greater.*

Proof Suppose we have two sequences, u and v, from an ACH system of period T. Let $k_{u,h}$ denote the number of timeslots in sequence u that are assigned with the channel index $h \in [0, N-1]$. Then, we can express the period length T as

$$T = \sum_{h=0}^{N-1} k_{u,h} = \sum_{h=0}^{N-1} k_{v,h} = \sum_{h=0}^{N-1} \left(\frac{k_{u,h} + k_{v,h}}{2} \right). \tag{3.2}$$

Without loss of generality, we fix u and cyclically rotate v by l, $l = 0, 1, \ldots, T-1$. Let us count the total accumulated number of rendezvous between u and $rotate(v, l)$ as l is incremented from 0 to $T - 1$. Since $C(u, rotate(v, l)) = N$, for any channel $h \in [0, N-1]$, the total number of rendezvous that involve a given timeslot, $v_i = h$, in v is $k_{u,h}$. Since there are $k_{v,h}$ timeslots in v that are assigned channel h, the total accumulated number of rendezvous between u and $rotate(v, l)$, as l is incremented from 0 to $T - 1$, in which the rendezvous channel is h is $k_{u,h} \cdot k_{v,h}$.

By definition, $C(u, rotate(v, l)) = N$ for any amount of cyclic rotation to v. This means that u and $rotate(v, l)$ must rendezvous in channel h at least once. Hence, the total accumulated number of rendezvous in channel h (as l is incremented from 0 to $T - 1$) is at least T. Thus, we have $k_{u,h} \cdot k_{v,h} \geq T$.

Since $\left(\frac{k_{u,h}+k_{v,h}}{2}\right)^2 \geq k_{u,h} \cdot k_{v,h}$, we can readily derive

$$\frac{k_{u,h} + k_{v,h}}{2} \geq \sqrt{T}. \tag{3.3}$$

Combining (3.2) and (3.3), we get

$$T = \sum_{h=0}^{N-1} \left(\frac{k_{u,h} + k_{v,h}}{2}\right) \geq N\sqrt{T}.$$

Therefore, we conclude that $T \geq N^2$. $\square$

In the next two sections, we describe two optimal ACH designs in which the period of the CH sequences is $O(N^2)$.

3.6.4 Asymmetric Optimal Asynchronous Channel Hopping Systems

Optimal ACH designs can be categorized into two types: the asymmetric design and the symmetric design that both achieve *maximum* rendezvous diversity (i.e., maximum number of rendezvous channels) with a *minimum* period of a CH sequence.

An asymmetric design of the optimal ACH system can be constructed using Algorithm 3 with an $N \times N$ array-based quorum system. A pair of sequences in an optimal ACH system has N pairwise rendezvous channels and the sequence period is N^2 timeslots. Thus, an optimal ACH system has an MRP value of $1/N$. We refer to S and R used in the construction algorithm as the *sequence array of the sender* and the *sequence array of the receiver*, respectively. In the above methods, channel indexes are assigned to array elements in a column-wise manner by the sender, or a span-wise manner by the receiver. Hence, we call the sender's CH sequence u as a *column-based* sequence, and the receiver's CH sequence v as a *span-based* sequence. Figure 3.11 illustrates an example ACH system built using a 3×3 array when $N = 3$ and u^* defines the sender's CH sequence after a cyclic rotation by $(2 + \delta)$.

3.6.5 Symmetric Optimal Asynchronous Channel Hopping Systems

3.6.5.1 Limitations of Asymmetric Asynchronous Channel Hopping Systems

An ACH system created using the asymmetric design approach requires each radio to have a pre-assigned role as either a sender or a receiver. Although such a requirement may be acceptable in some scenarios (e.g., half-duplex communication systems or in Bluetooth pairing), it may not be acceptable in systems where a radio's role as

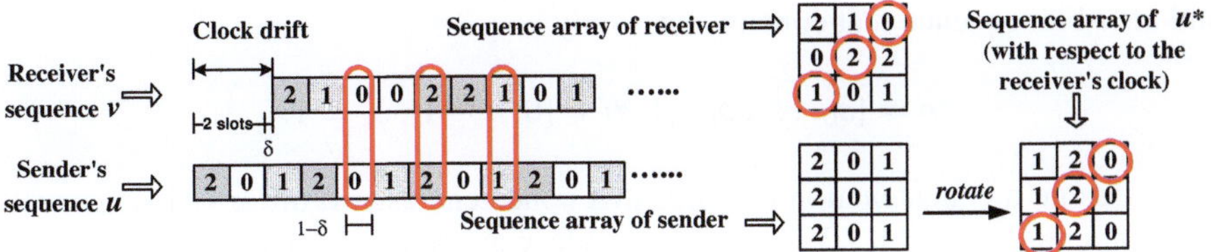

Fig. 3.11 An asymmetric design of an optimal ACH system when $N = 3$. The sender's clock is ahead of the receiver's clock by $(2 + \delta)$ timeslots where $\delta \leq 1/2$. The sequence arrays of the sender and the receiver that generated the respective CH sequences are shown. It can be easily seen that the two CH sequences have N distinct rendezvous channels despite the clock drift and that the overlap duration in each rendezvous channel is $(1 - \delta)$

sender/receiver is not pre-assigned. Designing time-asynchronous rendezvous protocols that are symmetric is a challenging problem. In the next subsection, we will explain how to devise symmetric ACH systems.

3.6.5.2 Construction of Symmetric Asynchronous Channel Hopping Systems

To construct a symmetric ACH system, we require every node to generate its CH sequence from a bit sequence that it possesses. Specifically, based on certain bit sequence design techniques, two channel hopping nodes are able to construct two bit sequences that have at least one distinct bit. Then, every node replaces any bit of "1" (or "0") in its bit sequence with a column-based (or span-based) sequence. The resulting CH sequences of the two nodes would achieve the maximum rendezvous diversity owing to the maximum number of rendezvous channels guaranteed between a column-based sequence and a span-based sequence. Note that every node follows the same method of constructing its CH sequence, which belongs to a symmetric ACH system.

We define a *bit sequence* as a sequence of bits where each bit takes either a value of 0 or 1. Note that a bit sequence is a special case of a CH sequence such that the channel indexes are chosen from $\{0, 1\}$. Hence, the cyclic rotation operation is also applicable to a bit sequence. Let us introduce the *bit sequence design problem* which is defined as follows: Enable two nodes to independently construct two *distinct* bit sequences. Note that the sequences resulting from cyclic rotations of a sequence are *not* considered to be distinct with respect to each other and the original sequence.

Assume that a *unique* n-bit sequence is assigned to each network node and that n is a system parameter. For example, the unique 48-bit ($n = 48$) sequence of a node can be the MAC address of the node's network interface. We refer to such a bit sequence as a node's *ID sequence*, and let α denote the n-bit ID sequence of node x. Now, let us define an n-bit sequence that contains zeros only:

$$z = \{z_0, \ldots, z_{n-1}\}, \forall i \in [0, n-1], z_i = 0,$$

and an n-bit sequence that contains ones only

$$o = \{o_0, \ldots, o_{n-1}\}, \forall i \in [0, n-1], o_i = 1.$$

Let $||$ denote a concatenation operator that concatenates two bit or CH sequences. We define the *expanded ID sequence* of node x as the concatenation of three n-bit sequences—α, z and o—as given in the following equation:

$$\mathbf{a} = \{a_0, \ldots, a_{3n-1}\} = \alpha||z||o.$$

According to the following lemma, two expanded ID sequences generated from two different ID sequences are guaranteed to be distinct.

Lemma 1 *Given any two n-bit sequences* $\alpha = \{\alpha_0, \ldots, \alpha_{n-1}\}$ *and* $\beta = \{\beta_0, \ldots, \beta_{n-1}\}$, *let* $\mathbf{a} = \alpha||z||o$ *and* $\mathbf{b} = \beta||z||o$, *where z is an n-bit sequence composed of only zeros and o is an n-bit sequence composed of only ones.*
   If $\alpha \neq \beta$, *then*
$$\mathbf{a} \neq rotate(\mathbf{b}, k), \forall k \in (0, 3n-1].$$

Proof We prove the above statement by considering three possible scenarios, and showing, in each scenario, that a bit in $\mathbf{a}$ and another bit in $rotate(\mathbf{b}, k)$ have different values although the two bits are in the same position within the respective expanded ID sequences. This is sufficient to prove that the two expanded ID sequences are not equal. Let $\mathbf{b}' = rotate(\mathbf{b}, k)$.
   Case 1: $k \in (0, n]$. Since $\mathbf{a} = \alpha||z||o$ and $\mathbf{b} = \beta||z||o$, $a_{2n-1} = 0$ and $b'_{2n-1} = 1$ (as indicated by the arrow in the Case 1 illustration of Fig. 3.12).
   Case 2: $k \in (n, 2n]$. Since $\mathbf{a} = \alpha||z||o$ and $\mathbf{b} = \beta||z||o$, $a_{3n-1} = 1$ and $b'_{3n-1} = 0$ (as indicated by the arrow in the Case 2 illustration of Fig. 3.12).
   Case 3: $k \in (2n, 3n-1]$. Since $\mathbf{a} = \alpha||z||o$ and $\mathbf{b} = \beta||z||o$, $a_{2n} = 1$ and $b'_{2n} = 0$ (as indicated by the arrow in the Case 3 illustration of Fig. 3.12).
   Thus, we conclude that $\mathbf{a} \neq rotate(\mathbf{b}, k), \forall k \in (0, 3n-1]$. $\square$

We now describe how to generate the CH sequences of a *symmetric* ACH system. Suppose that the number of available channels is N.

1. Suppose node x has a unique ID sequence, α, that contains n bits, and its expanded ID sequence is $\mathbf{a} = \alpha||z||o$, which has $3n$ bits. Lemma 1, ensures that two nodes that have two distinct ID sequences must have two distinct expanded ID sequences.

2. Using the procedure outlined in Sect. 3.6 for constructing asymmetric ACH systems, node x that needs to establish rendezvous with neighboring nodes generates its own asymmetric ACH system (including a column-based sequence u^x and a span-based sequence v^x) *independently* of other nodes. Note that different nodes may generate different asymmetric ACH systems.

3. Node x constructs its CH sequence by expanding each bit in $\mathbf{a}$ to a *frame*, which is mapped to a certain sequence as described below. From $i = 0$ to $i = 3n - 1$, each bit of $\mathbf{a}$ is expanded in the following way:

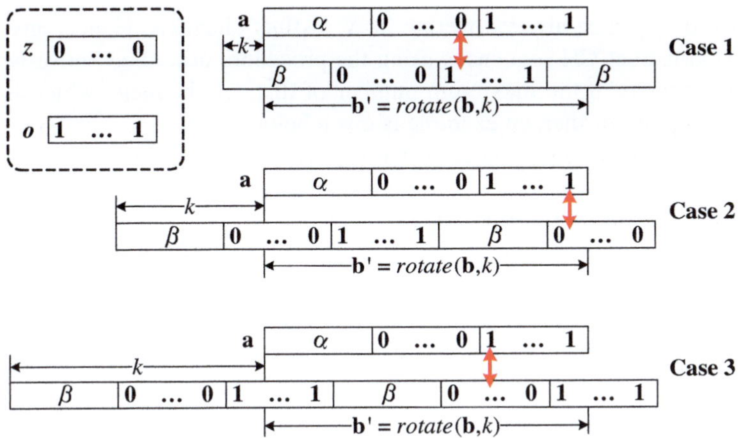

Fig. 3.12 Illustration of the three cases Lemma 1's proof

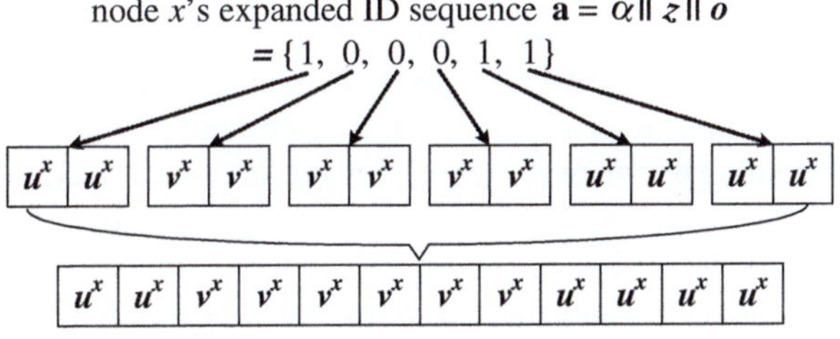

Fig. 3.13 Construction of a CH sequence for the symmetric ACH scheme. In the above figure, node x's ID sequence is $\alpha = \{1, 0\}$, $n = 2$, $z = \{0, 0\}$ and $o = \{1, 1\}$. Node x's column- and span-based sequences are u^x and v^x, respectively

(a) if $a_i = 1$, then the $(i+1)$th frame of node x's CH sequence is the sequence $(u^x || u^x)$;

(b) if $a_i = 0$, then the $(i+1)$th frame of node x's CH sequence is the sequence $(v^x || v^x)$.

Each frame is either a sequence $(u^x || u^x)$ or $(v^x || v^x)$, and thus it includes $2N^2$ slots.

A simple example of the CH sequence construction method is illustrated in Fig. 3.13. Every node has a distinct expanded ID sequence. In other words, every pair of *distinct* expanded ID sequences differ by at least one bit. For the sake of this discussion, let us say that they differ by a single bit at the i-th bit position. This implies that after the two expanded ID sequences are expanded, the i-th frame in each of the two resulting CH sequences will be composed of different types of sequences—viz, one frame is composed of a column-based sequence while the other frame is composed of a span-based sequence. As stated previously, a column-based sequence and a span-based sequence always overlap in N distinct channels. Hence, any pair of nodes that construct CH sequences using the procedure described above would be able to rendezvous in the maximum number of distinct channels, which is N. To illustrate this point further, an example is given below.

Suppose node x generates a column-based sequence u^x and a span-based sequence v^x. Likewise, node y generates u^y and v^y. Moreover, suppose nodes x and y have expanded ID sequences 100011 and 000011, respectively. Following the aforementioned procedure, the two nodes generate their CH sequences respectively as

$$(u^x||u^x)||(v^x||v^x)||(v^x||v^x)||(v^x||v^x)||(u^x||u^x)||(u^x||u^x)$$

and

$$(v^y||v^y)||(v^y||v^y)||(v^y||v^y)||(v^y||v^y)||(u^y||u^y)||(u^y||u^y).$$

The two nodes' expanded ID sequences are different in the first bit position, and the first frame of each node's CH sequence is $(u^x||u^x)$ and $(v^y||v^y)$ respectively. Since the first frame of one CH sequence is composed of two concatenated column-based sequences and the other is composed of two concatenated span-based sequences, nodes x and y are able rendezvous in N distinct channels during the first half frame.

Note that node x uses *two concatenated* column-based sequences, $(u^x||u^x)$, to expand a bit with a value of one or two concatenated span-based sequences, $(v^x||v^x)$, to expand a bit with a value of zero, instead of using a *single* column— or span-based sequence. This method is employed to guarantee that any pair of nodes that generate their CH sequences using the above procedure can rendezvous in N channels irrelevant of the clock drift between them. For instance, if an expanded ID sequence's bit is expanded to a single column (or span) based sequence, then rendezvous in N channels cannot be guaranteed when the clock drift is greater than zero and less than the length of a frame. An example is given in Fig. 3.14.

Algorithm 4 CH Sequence Construction for Symmetric ACH Systems.

Input: an n-bit sequence, α, and an asymmetric ACH system $\{u, v\}$ where u is a column-based sequence and v is a span-based sequence.
Output: a CH sequence w.

1: $\mathbf{a} = \alpha||z||o.$

2: **for** $i = 0$ to $3n - 1$ **do**
3:     **if** $a_i == 1$ **then**
4:         the $(i + 1)^{th}$ frame of w is $(u||u)$.
5:     **end if**
6:     **if** $a_i == 0$ **then**
7:         the $(i + 1)^{th}$ frame of w is $(v||v)$.
8:     **end if**
9: **end for**

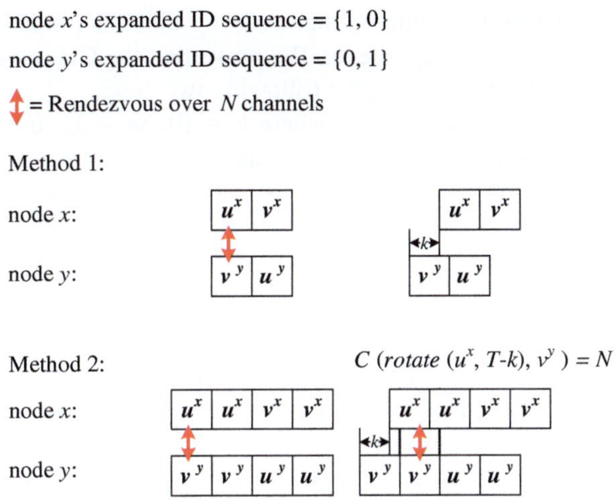

Fig. 3.14 Motivation for using two concatenated column/span-based sequences instead of a single sequence. Suppose node x generates column and span-based sequences u^x and v^x; similarly, node y generates sequences u^y and v^y. Using two concatenated column or span-based sequences in the CH sequence construction method (method 2 in the figure), the maximum rendezvous diversity is guaranteed when two channel hopping nodes have a clock drift that is less than a frame length. In the example, given N channels and a CH period of T, $rotate(u^x, T - k)$ yields another column-based sequence for node x, which overlaps with node y' span-based sequence v^y over N channels. On the contrary, the method of using a single column or span-based sequence (method 1 in the figure) is unable to guarantee such maximum rendezvous diversity, e.g., u^x may not have N rendezvous channels with a sequence concatenated by parts of v^y and u^y

Algorithm 4 describes the procedure for generating CH sequences of a symmetric ACH system. The following theorem describes the properties of the CH sequences generated using Algorithm 4.

Theorem 6 *Algorithm 4 generates sequences of an ACH system whose period is $6nN^2$ and degree of overlapping is N. Here, N is the total number of channels, and n is the length of a radio's ID sequence.*

Proof Suppose two arbitrary channel-hopping nodes, x and y, have n-bit ID sequences α and β, respectively. Let **a** and **b** be the expanded ID sequences of nodes x and y. Each node generates a column-based sequence and a span-based sequence *independently* of the other node. Suppose node x generates a column-based sequence u^x and a span-based sequence v^x. Similarly, node y generates a column-based sequence u^y and a span-based sequence v^y. Using its n-bit ID sequence, α, and the column-based sequence u^x and the span-based sequence v^x as inputs to Algorithm 4, node x generates its CH sequence w. Similarly, using its n-bit ID sequence, β, u^y and v^y as inputs to Algorithm 4, node y generates its CH sequence w'.

Since each CH sequence generated by Algorithm 4 (either w or w') contains $3n$ frames and each frame has exactly $2N^2$ slots, the period of a CH sequence generated by the algorithm is $6nN^2$.

Without loss of generality, suppose that node x's clock is i slots ahead of node y's clock, where i is an arbitrary integer. We prove that the CH sequences $\{w, w'\}$ form an ACH system by considering the following two cases.

Case 1: $2kN^2 < i \le (2k+1)N^2$, where $k \in [0, 3n-1]$. In this case, node x's CH sequence is ahead of node y's CH sequence by k frames and $(i - 2kN^2)$ slots. With respect to node y's clock, node x's CH sequence is equivalent to a CH sequence generated from the bit sequence $rotate(\mathbf{a}, \kappa)$. Let $\mathbf{a}' = rotate(\mathbf{a}, \kappa)$, where $\kappa = i - 2kN^2$. According to Lemma 1, we have $rotate(\mathbf{a}, \kappa) \ne \mathbf{b}$. If $a'_j \ne b_j$ for some $j \in [0, 3n-1]$, there are two possible cases to consider:

1. If $b_j = 0$, then $a'_j = 1$. This implies that the $(j+1)$-th frame of w' is $(v^y || v^y)$ and the $((j + \kappa + 1) \bmod 3n)$-th frame of w is $(u^x || u^x)$.
2. If $b_j = 1$, then $a'_j = 0$. This implies that the $(j+1)$-th frame of w' is $(u^y || u^y)$ and the $((j + \kappa + 1) \bmod 3n)$-th frame of w is $(v^x || v^x)$.

Case 2: $(2k+1)N^2 < i \le (2k+2)N^2$, where $k \in [0, 3n-1]$. In this case, node x's CH sequence is ahead of node y's CH sequence by an amount of $(k+1)$ frames and $(i - (2k+2)N^2)$ slots. With respect to node y's clock, node x's CH sequence is equivalent to a CH sequence generated from a bit sequence $rotate(\mathbf{a}, \kappa + 1)$. Let $\mathbf{a}' = rotate(\mathbf{a}, k + 1)$ and $\kappa = (2k+2)N^2 - i$. According to Lemma 1, we have $rotate(\mathbf{a}, \kappa + 1) \ne \mathbf{b}$. If $a'_j \ne b_j$ for some $j \in [0, 3n-1]$, there are also two possible cases to consider:

1. If $b_j = 0$, then $a'_j = 1$. This implies that the $(j+1)$-th frame of the CH sequence w' is $(v^y || v^y)$, the $((j + \kappa + 2) \bmod 3n)$-th frame of CH sequence w is $(u^x || u^x)$.
2. If $b_j = 1$, then $a'_j = 0$. This implies that the $(j+1)$-th frame of the CH sequence w' is $(u^y || u^y)$, the $((j + k + 2) \bmod 3n)$-th frame of CH sequence w is $(v^x || v^x)$.

In both cases, according to Theorem 2, a pair of sequences $(v^y || v^y)$ and $(u^x || u^x)$ (or $(u^y || u^y)$ and $(v^x || v^x)$) form an ACH system that has a degree of overlapping of N. This means that the pair of sequences have N distinct rendezvous channels irrelevant of the κ-slot clock drift when $0 \le \kappa \le N^2$. Thus, w and w' have N distinct rendezvous channels.

Since i is an arbitrary value, we can conclude that $\forall k, l \in [0, T-1]$,

$$C(rotate(w, k), rotate(w', l)) = N,$$

and $\{w, w'\}$ is an ACH system whose degree of overlapping is N. $\qquad\square$

According to Theorem 6, each node can rendezvous with another node without clock synchronization by independently generating ACH sequences using Algorithm 4. These ACH sequences enable maximum rendezvous diversity (i.e., rendezvous in N distinct channels within a period of $6nN^2$). Hence, the resulting ACH system's MRP value is $1/6nN$, and its ATTR is $O(N)$.

Note that no two distinct nodes will generate the same CH sequence since their expanded ID sequences are different. Using Algorithm 4, any two nodes can achieve asynchronous rendezvous without being pre-assigned a sender/receiver role.

Table 3.2 A comparison of asynchronous CH schemes

	Degree of overlapping	MRP	ATTR
RCH	0	n/a	N
SR	1	$\frac{1}{N(N+1)}$	$O(N^2)$
Asym. opt. ACH	N	$\frac{1}{N}$	$O(N)$
Sym. opt. ACH	N	$\frac{1}{6nN}$	$O(N)$

3.6.6 Comparisons

In this section, we compare the proposed optimal ACH designs with existing CH
schemes that also support the asynchronous operation. We assume that N is the total
number of available channels.

Random channel hopping (RCH) [91]. In this scheme, each node hops from one
channel to another randomly at the beginning of every timeslot. The RCH scheme
has an ATTR value of N. However, the RCH scheme does not guarantee a bounded
TTR between any two sequences, and its MRP value cannot be calculated either.

Sequence-based rendezvous (SR) [23]. Every node using SR follows the same
pre-determined CH sequence. The sequence period is $N(N + 1)$, the degree of
overlapping is 1, and thus the MRP is $\frac{1}{N(N+1)}$.

A comparison of all the aforementioned schemes is summarized in Table 3.2. From
the table, we can see that the proposed asymmetric and symmetric ACH schemes are
superior to the other schemes in terms of degree of overlapping. Both schemes' degree
of overlapping is the maximum possible value of N, and therefore the two schemes are
optimal in terms of rendezvous diversity. On the other hand, the two schemes' ATTR
is $O(N)$, which indicates that the ATTR scales linearly with the number of channels.
The proposed optimal ACH systems and the SR scheme guarantee a bounded TTR
between any two channel hopping nodes, while the RCH scheme cannot give an
upper bounded TTR.

3.6.7 Performance Evaluation

3.6.7.1 Simulation Setup

In this section, we compare the performance of three *symmetric* CH schemes that
do not require clock synchronization, namely RCH, SR, and the symmetric optimal
ACH, via simulation results. We simulate ten pairs of secondary nodes in a $1000\,\text{m} \times
1000\,\text{m}$ square area. Every secondary node has a single half-duplex radio, and the
transmission range of every secondary radio is $250\,\text{m}$. In the simulations, the radios of
all the secondary users and the primary users can access 11 channels (i.e., the number
of channels available to the CR network is $N = 11$). The duration of a timeslot is

10 ms. Each secondary node independently generates its CH sequence using the agreed CH scheme and performs channel hopping in accordance with the sequence. Once two nodes rendezvous on a channel, the link between them is established. Ten independent simulation runs were conducted for each simulation result.

Primary user (PU) traffic generation. We simulated X primary transmitters operating on X channels, and those channels were randomly chosen in each simulation run. In most existing work, it is assumed that a primary transmitter follows a "busy/idle" transmission pattern on a licensed channel [44], and we make the same assumption. The busy period has a fixed length of b timeslots, and the idle period follows an exponential distribution with a mean of l timeslots. All of the secondary nodes are within the transmission range of any primary transmitter. A channel is considered "unavailable" when primary user signals are present in it. All secondary nodes should refrain from transmitting in unavailable channels.

Random clock drift. The simulations were performed under the assumption that the nodes' clocks are not synchronized. In each simulation run, each node determines its clock time independently of other nodes. Thus, there is a random clock drift between any two nodes.

3.6.7.2 Impact of Degree of Overlapping

To show the impact of degree of overlapping on the rendezvous process, we first simulate a scenario in which primary transmitters continuously transmit on X channels, by setting b equal to the simulation time (or $b = \infty$) without any idle period.

Proportion of rendezvous pairs. We use the term *rendezvous pair* to denote a pair of nodes that are able to rendezvous successfully. We also define the term *proportion of rendezvous pairs* as the ratio between the number of rendezvous pairs and the total number of node pairs that attempt rendezvous expressed as a percentage. We measured the proportion of rendezvous pairs while varying the number of primary transmitters, and the results are shown in Fig. 3.15. As expected, SR showed inferior performance compared to the other two schemes due to the fact that they have small degree-of-overlapping values (see Table 3.2). When the degree of overlapping of a CH scheme is small, the scheme is more vulnerable against rendezvous failures caused by the presence of PU signals. In contrast, the optimal ACH scheme has the maximum possible value for the degree of overlapping, viz N, which explains its robustness against the aforementioned vulnerability.

ATTR. In Fig. 3.16, we compare optimal ACH and RCH in terms of ATTR. Although RCH has a high proportion of rendezvous pairs (see Fig. 3.15), it requires a significantly greater TTR, compared to optimal ACH, to establish rendezvous.

Rendezvous rate. We define the *rendezvous rate* as the number of successful rendezvous per slot. The rendezvous rate is another measure for quantifying a CH scheme's ability to establish rendezvous in the presence of PU signals. The measured rendezvous rates for the three CH schemes are given in Fig. 3.17. Note that MRP is a theoretical estimate of the rendezvous rate when all of the N channels are free of primary user signals. The rendezvous rate is an empirically-obtained measure of

Fig. 3.15 The proportion of rendezvous pairs versus number of primary transmitters

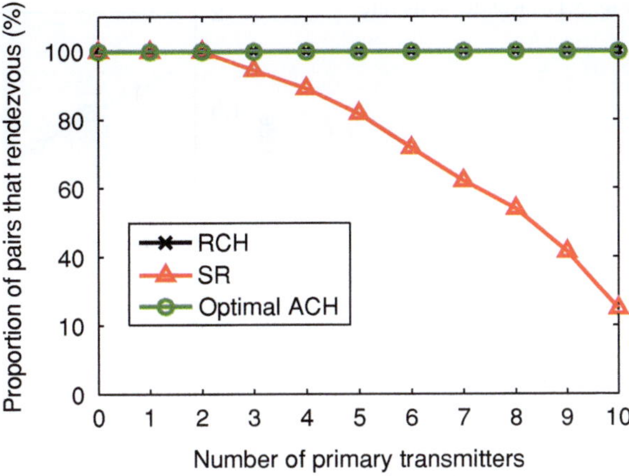

Fig. 3.16 Average TTR versus number of primary transmitters

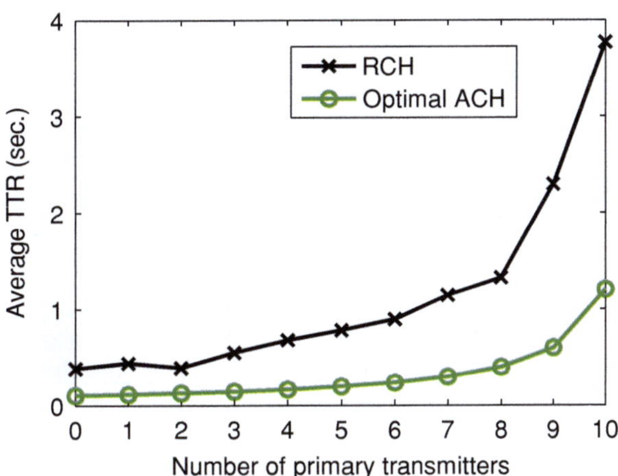

the number of rendezvous per slot in the presence of PU transmissions. Figure 3.17 shows that the optimal ACH scheme has the highest rendezvous rate among the three CH schemes.

3.6.7.3 Impact of Dynamic Primary User Traffic

In this set of simulations, we compare the three CH schemes when the primary user transmission parameters vary dynamically. Specifically, two parameters are varied: the length of the busy period, b, and the mean of the idle period, l.

In Figs. 3.18 and 3.19, we show the average TTR of the three schemes as l is varied while fixing parameters X and b. From both of the figures, we can observe that as the spectrum availability of each channel ($\frac{l}{l+b}$) increases, the measured ATTR

Fig. 3.17 Rendezvous rate

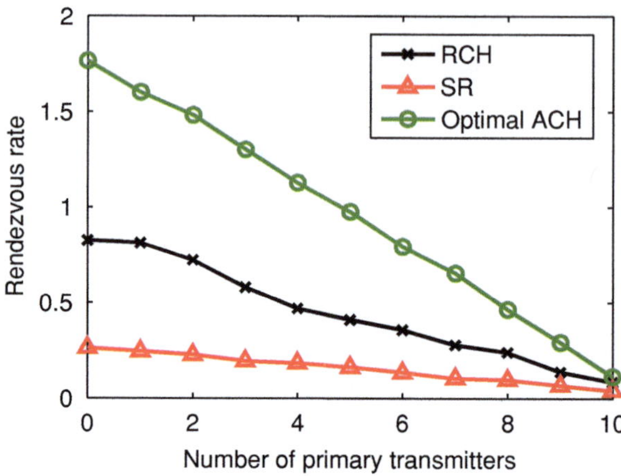

Fig. 3.18 Average TTR
versus mean of PU's idle
period when $b = 10$

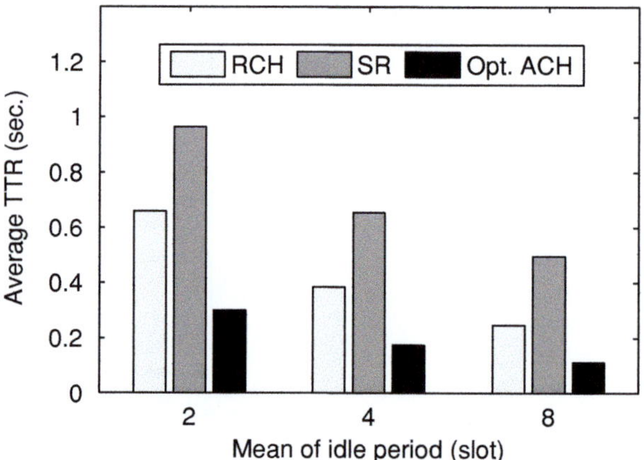

values of all CH schemes decrease. Note that the performance of SR is inferior to
that of RCH and optimal ACH due to its long period length ($N(N + 1)$) and low
degree of overlapping value. From Fig. 3.19, we can observe that the increase in l,
from $l = 2$ to $l = 8$, while fixing $b = 1$ has little effect on decreasing the average
TTR of the three CH schemes. This phenomenon can be attributed to the fact that the
availability of the channels is already sufficiently high when $l = 2$, which enables
most node pairs to readily rendezvous without encountering primary user-occupied
channels.

Based on our analytical and empirical results, we can make the following
conclusions about the relative performance of the asynchronous CH schemes under
consideration: the CH schemes that have maximum rendezvous diversity (e.g., optimal ACH) or provide rendezvous opportunities on every channel (e.g., RCH) would
outperform the CH scheme that has a limited degree of overlapping value but a long
period length (e.g., SR), in terms of the average TTR.

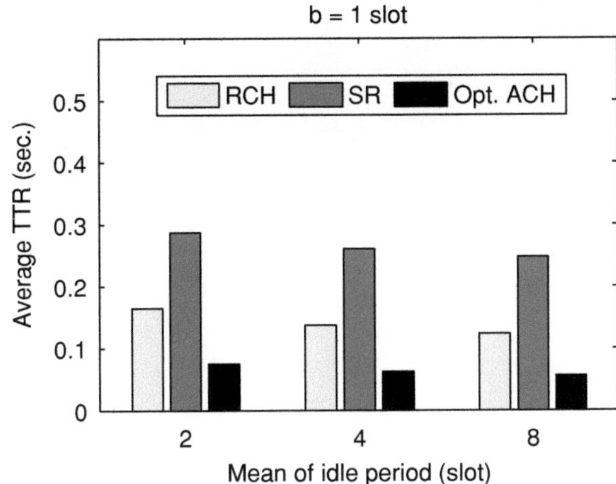

Fig. 3.19 Average TTR versus mean of PU's idle period when $b = 1$

3.7 Summary

In this chapter, we presented a systematic approach, based on *quorum systems*, for designing and analyzing channel hopping (CH) protocols that enable the rendezvous (or control) channel establishment in CR networks. A noteworthy feature of the proposed *Quorum-based Channel Hopping* (QCH) system is that it can establish rendezvous (or control) channels in multiple frequency channels so that the secondary network is less vulnerable to the unpredictable appearance of incumbent signals. Our approach is novel in that it considers the maximum degree of overlapping, clock synchronization and channel access latency issues in the design of asynchronous CH-based rendezvous protocols. There are two fundamental challenges in devising such a CH system: (1) establish pairwise rendezvous between two CH sequences of the sender and the receiver on *every* available channel; and (2) ensure that two asynchronous CH sequences achieve rendezvous with an *upper bounded* time-to-rendezvous (TTR).

We proposed two synchronous designs of the QCH system: the first design minimizes the MTTR of the CH system and the second design guarantees the even distribution of the rendezvous points in terms of both time and frequency. Minimizing the MTTR ensures short expected TTR which decreases channel access delay. An even distribution of rendezvous points alleviates the rendezvous convergence problem. Using the array-based quorum system, we have devised asynchronous CH schemes that meet the two challenges above. We have shown, using analytical and simulation results, that the CH schemes designed using the quorum-based framework outperform existing schemes under various network conditions.

Chapter 4
Coexistence-Aware Spectrum Sharing for Homogeneous Cognitive Radio Networks

Horizontal coexistence is one of the two representative coexistence problems in wireless networks, which is further divided into homogeneous and heterogeneous coexistence, as introduced in Chap. 1. In this chapter, we propose an inter-BS *Coexistence-aware Spectrum Sharing (CASS)* protocol for horizontal coexistence of homogeneous CR networks. CASS has the following noteworthy features: (1) it supports both non-exclusive and exclusive spectrum sharing and can dynamically switch between the two to minimize self-interference while keeping control overhead (induced by channel contentions) under control; (2) it uses a new channel selection algorithm that utilizes spectrum sensing results in order to minimize the likelihood of interference to incumbent transmissions; and (3) it uses an inter-BS channel contention procedure that enables a BS in need of more channels to borrow channels from its neighboring cells more readily (compared to 802.22's ODSC). Our analytical and simulation results show that these proposed schemes can effectively address the aforementioned MAC layer coexistence problems in CR networks.

In Sect. 4.1, we introduce the homogeneous coexistence mechanisms prescribed in IEEE 802.22 CR networks, and proposed by the research community. In Sect.4.2, we propose the Coexistence-aware Spectrum Sharing (CASS) protocol, and we evaluate the performance of CASS in Sect. 4.3. We conclude this chapter in Sect. 4.4.

4.1 Homogeneous Coexistence Mechanisms in IEEE 802.22 Networks

4.1.1 PHY-Layer Support and MAC-Layer Control Messages in IEEE 802.22

802.22 employs a Distributed Spectrum Sensing (DSS) framework to identify fallow licensed bands that are free of incumbent signals. In a cell, every CPE is required to report its local spectrum sensing results to its BS via intra-cell measurement

© Springer International Publishing Switzerland 2014
K. Bian et al., *Cognitive Radio Networks*,
DOI: 10.1007/978-3-319-07329-3_4

(control) messages. The BS has multiple radio interfaces for monitoring a number of channels simultaneously, and it determines the existence of incumbent signals on those channels using the local spectrum sensing reports from CPEs.

802.22 also defines inter-cell control messages as *beacons* to support the inter-BS communication. In every 802.22 time frame, there is a self-coexistence window for a BS to exchange beacons with neighboring BSs on multiple channels. If a beacon from a neighboring BS is received on a channel, that neighboring BS is deemed as a *co-channel BS* by the receiver BS. When direct inter-BS communication is infeasible, a BS instructs its associated CPEs (a.k.a. *bridge CPEs*) to listen for beacons from overlapping cells and report the information back. This feature enables a BS to obtain accurate information about other cells.

4.1.2 Non-exclusive Spectrum Sharing: Resource Renting

In an inter-BS dynamic resource sharing process, 802.22 defines *offerers* to be the BSs that have available resources and the *renter* to be the BS in need of spectrum. A renter BS initiates a Resource Renting process by broadcasting a *resource request* using inter-cell beacons. Upon reception of the resource request, neighboring BSs respond by broadcasting beacons that indicate their active channel sets and candidate channel sets. The *union* of candidate channel sets from neighboring BSs forms the renter BS's *grand candidate channel set* from which it selects channels to rent. The renter BS sends beacons to the offerers indicating the selected channel number and the duration of renting time. The Resource Renting process is concluded when the offerers transmit an acknowledgement back to the renter. 802.22 prescribes the following rule that the renter needs to comply with when selecting a channel:

Rule 1 *The renter should select a channel that interferes with the minimum number of channels being used by its neighbors.*

After selecting a channel, the renter BS has to determine whether non-exclusive sharing of the selected channel is feasible using the following rule:

Rule 2 *Non-exclusive spectrum sharing is feasible as long as the maximum achievable signal-to-interference ratio (SIR) on the selected channel is higher than the required SIR threshold of the network's supported services.*

If non-exclusive sharing is feasible, the BS schedules data transmissions on the selected channel with appropriate transmission power control settings.

4.1.2.1 Self-interference Caused by Resource Renting

The above Resource Renting mechanism may cause severe self-interference among overlapping cells. We use $r(x)$ to denote the minimum number of channels required

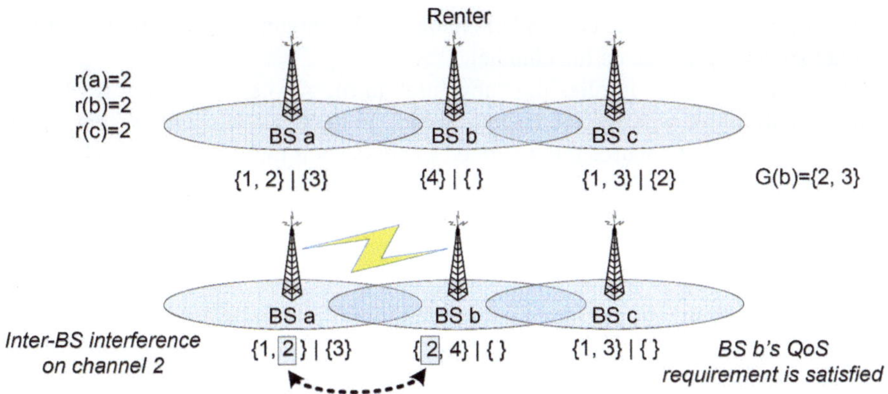

Fig. 4.1   Inter-cell resource renting in 802.22

by a BS x to satisfy the QoS of its workload. Let $G(x)$ denote the grand candidate channel set for the renter BS x. In Fig. 4.1, we illustrate a scenario with three overlapping cells a, b, and c. In this example, we have four channels in total, and $r(a) = r(b) = r(c) = 2$. BS a's active channel set is $\{1, 2\}$ and its candidate channel set is $\{3\}$. BS c's active channel set is $\{1, 3\}$ and its candidate channel set is $\{2\}$. Since BS b only has one active channel 4, it needs more channels to satisfy the QoS of its workload. Its neighboring BSs, a and c, have candidate channels to share with BS b. According to Rule 1, BS b can select channel 2 in $G(b)$ to rent from BS c. Unfortunately, BS b's renting of channel 2 will cause interference to BS a's communications since BS a also uses channel 2. In this case, self-interference between BS a and BS b may severely degrade the network performance of both cells.

4.1.3 Exclusive Spectrum Sharing

To completely avoid self-interference, 802.22 also prescribes exclusive spectrum sharing of the selected channel (target channel) via ODSC [38]. The renter BS becomes the contention source when initiating the ODSC. The contention source randomly selects a *channel contention number* (CCN) that is uniformly distributed in the range $[0, W - 1]$, where W is a constant representing the contention window size. The contention source includes its CCN and the target channel number in a spectrum contention request that it broadcasts to contention destination BSs. After receiving a spectrum contention request, a contention destination selects a CCN in the same manner as the contention source. Then the contention destination determines the outcome of the pair-wise contention using this contention resolution rule: the BS with a greater CCN value wins the pair-wise contention. This rule implies that the contention source's probability of winning a pair-wise contention is $1/2$. If the contention source wins the contended channel, all contention destinations perform channel switching to vacate the channel.

This contention rule implies the "absolute" fairness—i.e., that every BS has an equal probability of winning the target channel. In Sect. 4.2, we will present a contention protocol that ensures the weighted fairness such that a BS with higher QoS requirement or traffic load will have a greater probability of winning the target channel.

4.1.4 Other Related Work

In concept, 802.22's non-exclusive spectrum sharing mechanisms are somewhat similar to channel borrowing schemes proposed for cellular telephone networks [55]. In a channel borrowing scheme, a BS of one cell borrows channels from adjacent BSs. However, there is one important difference between 802.22's non-exclusive spectrum sharing and channel borrowing in cellular telephone networks—the former is a distributed process whereas the latter is a centralized process. 802.22 cells will be likely managed by different wireless service providers, and thus there will be no central entity that has the authority to manage channel assignments between 802.22 cells. For this reason, 802.22 carries out spectrum sharing (through channel borrowing) in a distributed manner. In contrast, channel assignment and sharing in cellular telephone systems are carried out in a centralized manner since cells in a given region are typically controlled by a centralized base station controller.

The research community's interest in the air interface for dynamic spectrum access has intensified in recent years. In [87], Sengupta et al. propose improvements for 802.22's air interface. Specifically, they formulated the 802.22 channel assignment problem as a vertex coloring problem, and proposed a coloring algorithm called Utility Graph Coloring (UGC) that is a centralized algorithm designed to maximize system utility (i.e., spectrum reuse). In [44], the authors studied techniques to maximize throughput in dynamic spectrum access networks under incumbent protection constraints.

4.2 Protocol Descriptions

4.2.1 Basic Assumptions

4.2.1.1 Incumbent Traffic

We assume that the presence or absence of incumbent users' signals on a channel, say channel i, can be modeled as a continuous-time "ON/OFF" Poisson process

[44], where inter-arrival times of consecutive incumbent signals are exponentially-distributed with a rate parameter. Let the random variable V_i denote the length of an incumbent's idle interval (*incumbent idle time period*) on channel i. Similarly, let the random variable U_i denote the length of an incumbent's busy interval (*incumbent busy time period*) on channel i. Suppose that $E[V_i] = v_i$ and $E[U_i] = u_i$. As shown in [59], the value of v_i and u_i can be estimated using spectrum sensing results. The probability that channel i is free of incumbent users' signals is

$$\alpha_i = \frac{v_i}{v_i + u_i}.$$

4.2.1.2 QoS Requirement

The spectrum contention process would require the coexisting networks to exchange potentially sensitive information—such as QoS requirements, traffic load, and network characteristics—to negotiate and resolve the contention among them. We need a measure to quantify the traffic load of a coexisting network. For this purpose, we use $r(x)$ to denote the minimum number of channels required by a BS x to satisfy the QoS of its workload.

4.2.1.3 Inter-BS Communication

802.22 defines inter-BS communication methods: over-the-air and over-the-backhaul in support of the inter-BS communication. These features enable a BS to exchange control information with neighboring BSs in a timely and efficient manner.

4.2.2 *Dynamic Switching Between the Two Spectrum Sharing Modes*

CASS supports two spectrum sharing modes: non-exclusive spectrum sharing and exclusive spectrum sharing. It can switch from one mode to the other depending on the channel conditions. CASS performs channel evaluation to determine when to switch from one mode to the other. In this channel evaluation process, CASS needs to calculate the channel capacity in each of the two spectrum sharing modes.

4.2.2.1 Channel Capacity in Non-exclusive Spectrum Sharing Mode

If the renter BS shares channel i with its neighbors, self-interference may cause packet collisions. Let $\gamma(i)$ denote the maximum achievable SIR perceived by the renter on the selected channel i. Let $e(i)$ denote the packet error rate on the selected channel

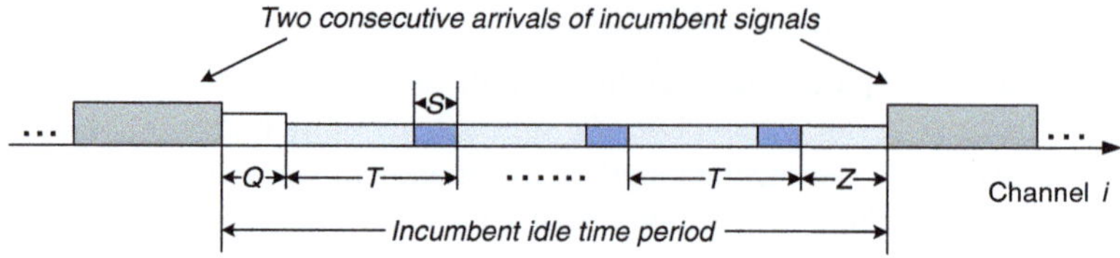

Fig. 4.2 Incumbent idle time period and channel contention periods

at a receiver (a CPE) in the renter's cell. The capacity of channel i in non-exclusive spectrum sharing can be expressed as

$$C_N(i) = \alpha_i \cdot (1 - e(i)) .$$

In [90], Shellhammer describes a way of estimating $e(i)$ based on $\gamma(i)$: given the modulation method, the symbol error rate (SER) can be estimated based on $\gamma(i)$; then, $e(i)$ can be calculated based on the SER.

4.2.2.2 Channel Capacity in Exclusive Spectrum Sharing Mode

If exclusive spectrum sharing of the selected channel is necessary, a contention phase is periodically scheduled so that the channel contention procedure can be invoked. Let T denote the duration of a channel contention period. At the end of every contention period, a contention phase of length S is scheduled, as shown in Fig. 4.2. S must be long enough to perform channel contention-related operations, such as determining the target channel(s), performing channel contentions (via the exchange of inter-BS beacons), and preparing to resume transmissions on the new channels (if new channels have been acquired or if channel switches have occurred).

The capacity of channel i in exclusive spectrum sharing can be calculated as

$$C_E(i) = \alpha_i \cdot \frac{T_X(i)}{v_i},$$

where $T_X(i)$ is the expected maximum transmission time for 802.22 entities on channel i during an incumbent idle time period.

802.22 entities cannot immediately start to access a channel once spectrum sensing determines that the channel is fallow. A quiet period of length Q is needed for performing spectrum sensing and channel setup before an 802.22 entity can access the channel, as shown in Fig. 4.2. Let the random variable τ and random variable L_i denote the length of the incumbent idle time period and the length of an 802.22 entity's maximum transmission time on channel i (during an incumbent idle time period), respectively. One of two possible scenarios can occur:

- if $\tau \leq Q$, $L_i = 0$;
- If $\tau > Q$, the incumbent idle time period is composed of a quiet period, $\lfloor \frac{\tau - Q}{T} \rfloor$ contention periods and a residual time interval Z as shown in Fig. 4.2, where $0 \leq Z < T$.

Let p_{FN} denote the false negative[1] probability of spectrum sensing in the contention source's cell. Now we can calculate the expected maximum transmission time on channel i as

$$T_X(i) = E[L_i] = 0 + (1 - p_{FN}) \cdot$$
$$\int_Q^\infty \left(\lfloor \frac{\tau - Q}{T} \rfloor \cdot (T - S) + \min(Z, T - S) \right) \cdot p_\tau(i) d\tau,$$

where $p_\tau(i)$ is the probability that there is no incumbent signal arrival in channel i during the time duration τ. This value can be calculated using the following equation:

$$p_\tau(i) = \int_\tau^\infty \frac{1}{v_i} \cdot e^{-\frac{u}{v_i}} du = e^{-\frac{\tau}{v_i}} .$$

4.2.2.3 An Approach of Switching Between Two Modes

In 802.22, the non-exclusive spectrum sharing of the selected channel is attempted first. As stated in Rule 2, if the maximum achievable SIR on the selected channel is not sufficiently high, 802.22 prescribes exclusive spectrum sharing. However, 802.22 fails to give specific guidelines on when to use exclusive sharing. We propose one possible approach for determining when to use exclusive spectrum sharing. CASS determines which spectrum sharing mode to operate in, for a given channel i, by comparing $\gamma(i)$ with the required SIR threshold parameter $\gamma^*(i)$. To obtain the value of $\gamma^*(i)$, we need to first solve for $e(i)$ in the equation $C_N(i) = C_E(i)$; and then find the SIR value corresponding to the solved $e(i)$, which is used as the value of $\gamma^*(i)$. CASS determines the spectrum sharing mode as follows:

1. If $\gamma(i) \geq \gamma^*(i)$, CASS operates in the non-exclusive spectrum sharing mode. The benefits of non-exclusive sharing (e.g., low control overhead) outweigh the benefits of exclusive sharing (e.g., no self-interference).
2. If $\gamma(i) < \gamma^*(i)$, CASS operates in the exclusive spectrum sharing mode. The benefits of exclusive sharing outweigh the benefits of non-exclusive sharing.

---

[1] This type of error occurs when the spectrum sensing results erroneously indicate that a channel is free of incumbent signals when in fact the opposite is true.

4.2.3 The Channel Selection Mechanism

The channel selection criterion in 802.22's mechanisms (Rule 1) implies that the renter BS does not consider any incumbent coexistence issues. It may not be suitable in environments where the appearance of incumbent signals are frequent and unpredictable—such a scenario would likely occur in areas where Part 74 devices operate. After the incumbent signals on active channels have been detected in a cell, the BS and CPEs need to vacate those channels as soon as possible and find suitable replacement channels. The incumbent users in the cell's vicinity are also likely to experience some level of interference. Instances of such undesirable situations can be reduced by considering the incumbent coexistence requirement during the channel selection process. We propose to use an incumbent protection criteria—probability of interfering with incumbent signals on an 802.22 channel. CASS prescribes that the renter BS should select an 802.22 channel that has the lowest probability of interfering with incumbent signals.

For a BS x in need of spectrum, let $N(x)$ denote the set of neighboring BSs of x. Let $C(y)$ denote BS y's set of candidate channels. BS x's grand candidate channel set is

$$G(x) = \bigcup_{y \in N(x)} C(y).$$

For channel i and BS y, let r_i^y denote the probability that BS y schedules data transmissions on channel i in a given time interval. The value of r_i^y for any channel i is determined by the BS y itself. When $r_i^y = 0$, this means that channel i is a candidate channel for BS y or channel i is not allocated to BS y. Neighboring BSs exchange information on each other's r_i^y value using inter-cell beacons.[2]

We define the *channel accessibility* of channel i for the renter BS x, represented by notation $\varphi_x(i)$, as the probability that BS x can access channel i with no incumbent or self-interference in a given time interval. We can compute $\varphi_x(i)$ using the following equation:

$$\varphi_x(i) = \alpha_i \prod_{y \in N(x)} \left(1 - r_i^y\right).$$

We define the set of *accessible candidate channels* for the BS x as

$$R(x) = \{i \,\|\, \varphi_x(i) > \eta,\ i \in G(x)\}, \tag{4.1}$$

where η is an adjustable parameter.

To comply with CASS's channel selection rule, BS x selects channel s such that

---

[2] In 802.22, a *quiet period* in every 802.22 frame is used for 802.22 entities to exchange inter-cell beacons. Information about the cell's candidate channel sets and active channel sets is contained in the beacons.

$$s = \arg \min_{i \in R(x)} \{p_{int}(i)\}, \tag{4.2}$$

where $p_{int}(i)$ is the probability of interfering with incumbent signals on channel i. Let t denote the expected length of secondary user packets. If v_i is known, $p_{int}(i)$ can be calculated as

$$p_{int}(i) = 1 - \int_t^\infty \frac{1}{v_i} \cdot e^{-\frac{u}{v_i}} du = 1 - e^{-\frac{t}{v_i}}. \tag{4.3}$$

CASS's channel selection technique minimizes the probability of interfering with incumbent signals (as prescribed by (4.3)) subject to the constraint on channel accessibility (as prescribed by (4.1)).

4.2.4 The Channel Contention Procedure

In ODSC's CCN-based channel contention procedure, every BS has the equal probability of winning the target channel, which is $1/k$ (k is the number of BSs participating in the contention process). This implies that a contention source that has a high QoS requirement is unlikely to win the needed channels when there are a large number of competing BSs. To address this problem, instead of using CCN, CASS employs the *Contention Priority Number* (*CPN*) to resolve pair-wise contentions.

4.2.4.1 Contention Priority Number

Let $H(i)$ define set of BSs that contend for channel i, and let p_x denote the CPN of BS x. In CASS, every BS $x \in H(i)$ that contends for channel i will broadcast its QoS requirement $r(x)$ to neighboring BSs.

After collecting the QoS requirements of competing BSs, every BS is able to determine the contention window size W as follows.

$$W = \sum_{x \in H(i)} r(x).$$

Every BS $x \in H(i)$ selects its CPN p_x from the range $[0, W-1]$ at random.

4.2.4.2 Token Assignment

After the contention windows size W is determined, let $\mathcal{K} = \{0, 1, 2 \ldots, W-1\}$ be a set that contains W tokens. Every BS $x \in H(i)$ picks a number of $r(x)$ tokens from $\mathcal{K}$ at random. Let K_x denote the set of tokens picked by BS x. As a result, all tokens in $\mathcal{K}$ are assigned to k competing BSs, i.e.,

$$\mathcal{K} = \bigcup_{x \in H(i)} K_x.$$

4.2.4.3 Contention Resolution Rule

CASS's contention resolution rule is straightforward: the winner BS is the one who has picked the winner token during the token assignment phase. Let ω denote the winner token number, and we introduce a simple algorithm that selects the winner token.

- Every BS $x \in H(i)$ broadcasts its selected CPN to all competing BSs.
- After collecting $(k-1)$ CPN values from other BSs, each BS is able to calculate the winner token as follows.

$$\omega = \sum_{x \in H(i)} p_x \quad (\mathrm{mod}\ W).$$

- The winner BS x^* is the one that has picked the token ω during the token assignment phase, i.e., $\omega \in K_{x^*}$.

The winner BS x^* adds channel i to its active channel set. The contention destination BS removes the contended channel i from its available channel list. Meanwhile, non-winner BSs that are involved in the channel contention process send beacons to its neighboring BSs to notify them about the result of the channel contention.

Next, we show that the above contention resolution rule guarantees the weighted fairness, i.e., the winning probability of BS x is proportional to its QoS requirement $r(x)$.

4.2.4.4 Weighted Fairness

Suppose BS $x \in H(i)$ has a QoS requirement $r(x)$. The CPN of BS x, p_x, is a random variable that uniformly distributed over the set $[0, W-1]$. Thus,

$$Pr\{p_x = v\} = \frac{1}{W}, \forall v \in [0, W-1].$$

Let S denote the summation of all CPNs of BSs in $H(i)$ except p_x, that is

$$S = \sum_{y \neq x, y \in H(i)} p_y.$$

Let $L = (p_x + S)$. Since

$$L = (p_x + S) \quad (\text{mod } W)$$
$$= p_x \quad (\text{mod } W) + S \quad (\text{mod } W),$$

then we have $\forall u \in [0, W-1]$,

$$Pr\{L = u\} = \sum_{v=0}^{W-1} (Pr\{p_x = v\} \cdot Pr\{S = (u-v) \quad (\text{mod } k)\})$$
$$= \sum_{v=0}^{W-1} \left(\frac{1}{W} \cdot Pr\{S = (u-v) \quad (\text{mod } k)\} \right)$$
$$= \frac{1}{W} \cdot \sum_{v=0}^{W-1} (Pr\{S = (u-v) \quad (\text{mod } k)\})$$
$$= \frac{1}{W}.$$

Since BS x has $r(x)$ tokens, the probability that one of BS x's picked tokens is the winner token is $\frac{r(x)}{W}$, which is also the probability that BS x wins the contention. As a result, BS x will have a higher winning probability than BS y if $r(x) > r(y)$.

4.3 Performance Evaluation

In this section, we compare CASS with 802.22's inter-BS resource sharing mechanisms via simulation results. All of the scenarios we considered involve multiple overlapping cells. In each cell, there is one BS and ten CPEs. Every BS requires five channels to satisfy the QoS of its admitted workload. BSs of overlapping cells are synchronized by the periodic transmission of inter-cell beacons. The simulation parameter values were chosen to be consistent with those used in the simulation experiments in [52], and they are given in Table 4.1. The 802.22 Working Group has suggested that a variation of the Hata model is the most appropriate propagation model for studying 802.22 [51]. The Hata model [85] represents the urban area propagation loss as a standard formula and supplies correction equations to model suburban and open rural areas. In our simulations, we use a variation of the Hata model for open rural areas. We simulated three inter-BS spectrum sharing protocols: 802.22 Resource Renting, 802.22 ODSC, and CASS. Note that Resource Renting is a pure non-exclusive spectrum sharing protocol; ODSC is a pure exclusive spectrum sharing protocol; and CASS can operate in either spectrum sharing mode, switching from one mode to the other when the channel conditions warrant it. We created our own event-driven simulator in C, and each simulation result is the average value of ten simulation runs.

Table 4.1 Default simulation parameter values

Simulation parameters	Values
Total licensed spectrum band	54–806 MHz
Bandwidth of a licensed channel	8 Mbps
Number of licensed channels	10
BS transmission radius	30 Km
TV transmission receiving radius	30 Km
Modulation method	QPSK
Effective antenna height of the BS	30 m
Effective antenna height of the CPE	9 m
Channel switch delay	10 ms
802.22 frame size	40 ms

We use the notation λ_i to denote the number of incumbent signal arrivals on channel i during one superframe,[3] assuming that the incumbent signals follow a Poisson arrival process—hereafter, we'll simply refer to this value as the incumbent transmission rate. We assume that there is at most one incumbent transmitter per channel per cell.

4.3.1 The 3-BS Scenario

We simulated a scenario with three overlapping 802.22 cells. The topography of the cells is shown in Fig. 4.1. The distance between two neighboring BSs is d Km. In this simulation experiment, the initial channel assignments are given as: BS a's active channel set is $\{1, 2, 3, 4, 9\}$ and its candidate channel set is $\{10\}$. BS c's active channel set is $\{1, 2, 3, 4, 10\}$ and its candidate channel set is $\{9\}$. BS b has only an active channel set $\{5, 6, 7, 8\}$, and it is in spectrum shortage. The grand candidate channel set of BS b is $G(b) = \{9, 10\}$. Let $\lambda_i = 0, i = 1, 2, \ldots, 9$, and $\lambda_{10} = 1/2$.

Figure 4.3 shows a plot of the throughput of BS b (renter) versus the distance between neighboring BSs, d. As expected, when d is small ($<$50 Km), CASS and ODSC outperformed Resource Renting. When the distance between neighboring BSs is small, the SIR measured at the 802.22 entities (BS and CPEs) of the renter cell is low due to self-interference. In such a situation, a non-exclusive spectrum sharing scheme, such as Resource Renting, would be expected to perform poorly. When $d = 50$ Km, Resource Renting's performance was comparable to that of ODSC because the effects of self-interference were insignificant at this distance.

Figure 4.3 shows that CASS outperforms both Resource Renting and ODSC irrelevant of the value of d. We can explain this result by recalling the fact that CASS attempts to select a channel that is most underutilized by the incumbent users by

---

[3] An 802.22 superframe is composed of 16 frames.

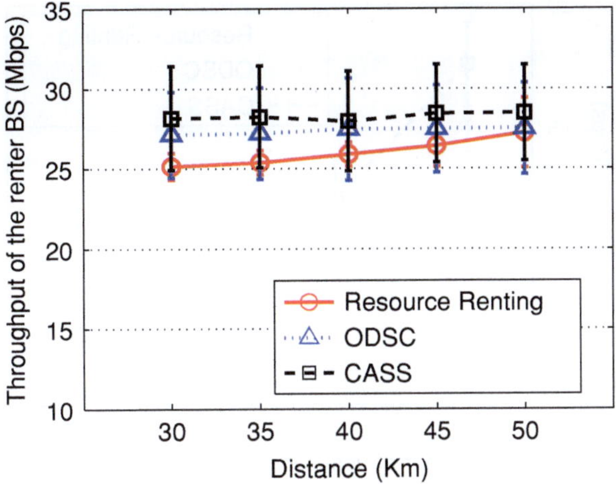

Fig. 4.3 Throughput of BS b versus distance d in the 3-BS scenario

requiring the renter BS to consider spectrum sensing results during the channel selection process; in contrast, the other two spectrum sharing schemes do not consider spectrum sensing results. When either Resource Renting or ODSC is used, the likelihood that an incumbent signal appears in the active channel(s) is higher compared to CASS. Obviously, the appearance of incumbent signals in an active channel degrades the renter cell's throughput as this requires costly channel switches. When CASS is used, BS b selects channel 9 from its grand candidate channel set because $\lambda_9 < \lambda_{10}$. In practice, a renter BS can use the following procedure to utilize spectrum sensing results when selecting a channel: (i) Using a compiled record of recent spectrum sensing data and a maximum likelihood estimator, estimate the expected incumbent idle time for each channel (see [59] for details); (ii) Using (4.3) and the values obtained in the first step, calculate the probability of interfering with incumbent signals for each channel; and (iii) Select a channel using (4.2).

4.3.2 The 9-BS Scenario

We simulated a network topography with 9 overlapping cells in a $100 \times 100\,\mathrm{Km}$ area. All channels are randomly assigned to the BSs. Figure 4.4 shows plots of average throughput per cell vs. the number of incumbent transmitters per cell. The incumbent transmission rate was fixed to $1/4$ for all incumbent transmitters. The performance gap is seen between Resource Renting and ODSC. In such a high self-interference environment, non-exclusive spectrum sharing schemes are inferior to exclusive spectrum sharing schemes. Figure 4.5 shows plots of average throughput per cell vs. incumbent transmission rate.

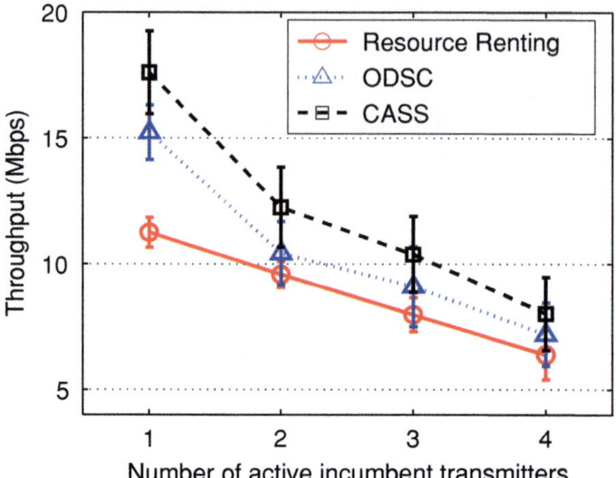

Fig. 4.4 Throughput versus Number of active incumbent transmitters in the 9-BS scenario

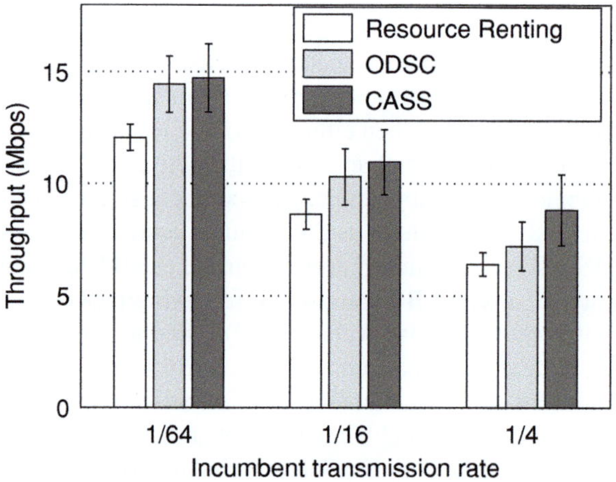

Fig. 4.5 Throughput versus Incumbent transmission rate in the 9-BS scenario

Figure 4.6 shows plots of the average throughput per cell vs. the channel switch delay. In the corresponding simulation experiments, we included four active incumbent transmitters that have the same transmission rate, viz 1/16. In inter-BS spectrum sharing, the channel switch delay is the biggest contributing factor to the control overhead. A recent study has found that the channel switch latency for most popular wireless cards is between 5 and 20 ms [72]. In Fig. 4.6, we used the same range of values for the channel switch delay. As the channel switch delay increases, the control overhead caused by channel switch events during the channel contention procedures increases, thereby degrading the performance of ODSC and CASS.

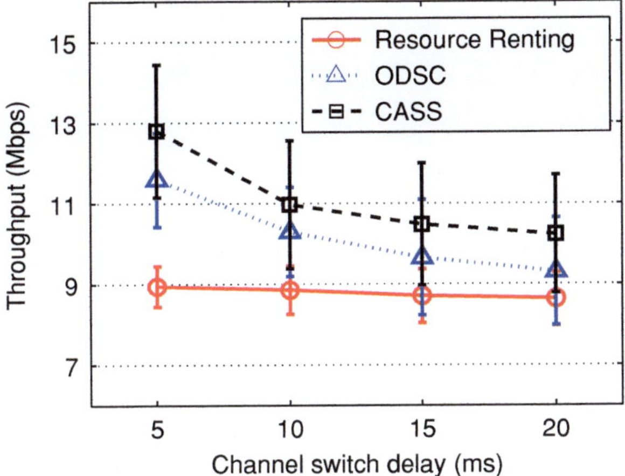

Fig. 4.6 Throughput versus Channel switch delay in the 9-BS scenario

4.4 Summary

The primary objective of 802.22 inter-BS spectrum sharing is to address the two major challenges in self-coexistence: minimize self-interference (among cells) and enable cells to acquire enough resources to satisfy the QoS of their admitted workloads (in opportunistic spectrum sharing environments). In this chapter, we identified the drawbacks of 802.22's spectrum sharing mechanisms and proposed a new inter-BS spectrum sharing protocol called *Coexistence-aware Spectrum Sharing (CASS)*. Our simulation results show that CASS outperforms 802.22's spectrum sharing mechanisms. The major contributions of this chapter include the following: (1) CASS gives us some level of insight on how to determine whether non-exclusive spectrum sharing or exclusive spectrum sharing should be employed; (2) CASS uses a channel selection strategy that incorporates spectrum sensing information to help satisfy incumbent coexistence requirements and improve network throughput; (3) CASS employs a channel contention procedure that was designed to avoid the drawbacks of ODSC's channel contention procedure.

Chapter 5
Frequency Reuse over a Single TV White Space Channel

By following the discussions on homogeneous coexistence in the previous chapter, we continue investigating this issue in this chapter, but in a slightly-different network environment where the available spectrum is insufficient to satisfy the QoS requirements of multiple CR networks and they may have to share a single TVWS channel. When multiple co-located CR networks operate in the same swath of whitespace (or unlicensed) spectrum with little or no direct coordination, co-channel self-coexistence is a challenging problem. In this chapter, we focus on the problem of spectrum sharing over a single TVWS channel among coexisting CR networks that employ orthogonal frequency division multiple access (OFDMA) in their uplink and do not rely on inter-network coordination. An uplink soft frequency reuse (USFR) technique is proposed to enable globally power-efficient and locally fair spectrum sharing.

In Sect. 5.1, we discuss the system model that underpins the proposed USFR technique as well as related issues. We frame the self-coexistence problem as a non-cooperative game, and the decoupled transmit power control (TPC) and subchannel allocation (SCA) subproblems are studied in Sects. 5.2 and 5.3, respectively. A two-level game-theoretic approach is proposed in Sect. 5.4 by integrating the SCA and TPC games as the Uplink Resource Allocation (URA) game. In Sect. 5.5, our simulation results show that USFR significantly improves self-coexistence in spectrum utilization, power consumption, and intra-cell fairness. We conclude this chapter in Sect. 5.6.

5.1 System Model

5.1.1 Uplink Soft Frequency Reuse

The idea of SFR proposed in the downlink is illustrated in Fig. 5.1a. In each of the co-channel cells, the user terminals next to their home BS, called inner users, can

© Springer International Publishing Switzerland 2014
K. Bian et al., *Cognitive Radio Networks*,
DOI: 10.1007/978-3-319-07329-3_5

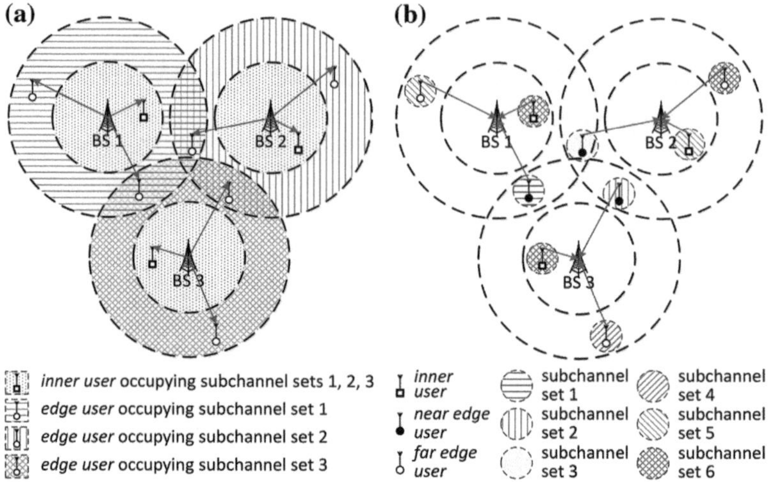

Fig. 5.1 Basic ideas: **a** SFR (*left*); **b** USFR (*right*)

fully occupy the entire common channel. But the user terminals far away from their home BS, called edge users, have to take an exclusive set of subchannels in each cell. Similarly, in Fig. 5.1b, the idea of USFR is to allow some users in a cell to share certain subchannels with some users in other cells. However, things are more complex in the uplink case. We still follow the same definitions of inner users and edge users. Furthermore, uplink edge users can be either *near* or *far* ones based on whether or not they are located close to any inter-cell overlapping areas. Hence, there can be a greater number of possible coexistence patterns, i.e., resource allocation patterns, created in the uplink by the combinations of various users.

5.1.2 Uplink Resource Allocation Problem

In each network cell, the BS conducts URA for the users under control. Particularly, local URA includes SCA and TPC for the users' uplink sessions. Whenever a new session becomes active in any cell, its home BS needs to redo URA to spare enough resource for the new session. Due to the dynamic interference environment where spectrum availability can be temporary and user terminals can be mobile/portable, we focus on a certain short time period, in which the interference environment is relatively static without the need for making prediction.

Suppose that there are total N cells coexisting on a common channel, which consists of K subchannels. In each cell n for $n \in \mathcal{N} \triangleq \{1, \cdots, N\}$, there are $M^{(n)}$ active sessions. Each user in cell n can maintain multiple sessions, and each session m_n for $m_n \in \mathcal{M}^{(n)} \triangleq \{1, \cdots, M^{(n)}\}$ can operate on multiple subchannels. We want to make the common channel accommodate all the uplink sessions that are active in the coexisting cells, but it is still possible that some sessions have to be blocked if the BS cannot find a feasible solution to the URA problem defined below. This issue is characterized by spectrum utilization. To avoid unnecessary intra-cell interference, each subchannel k for $k \in \mathcal{K} \triangleq \{1, \cdots, K\}$ cannot be assigned to more than one session in the same cell at the same time [70], since the sessions share the same home BS as a common receiver in the uplink. Because we focus on each short time period for the design of USFR, the same subchannel can still be allocated to multiple sessions in different time periods when USFR is carried out periodically.

In the cell $n = a$, local URA strategy is characterized by SCA and TPC strategy matrices, which are denoted by $\mathbf{U}^{(a)} \triangleq \{U^{(a)}_{m_a,k}\}_{M^{(a)} \times K}$ and $\mathbf{P}^{(a)} \triangleq \{P^{(a)}_{m_a,k}\}_{M^{(a)} \times K}$, respectively. Each $U^{(a)}_{m_a,k}$ in $\mathcal{U}^{(a)} \triangleq \{\mathbf{U}^{(a)}\}$ denotes a binary indicator that is equal to 1 (0) when session m_a takes (does not take) subchannel k, and each $P^{(a)}_{m_a,k}$ in $\mathcal{P}^{(a)} \triangleq \{\mathbf{P}^{(a)}\}$ denotes the corresponding allocated transmit power. Each pair of $U^{(a)}_{m_a,k}$ and $P^{(a)}_{m_a,k}$ has the following relationship: $P^{(a)}_{m_a,k} > 0$ when $U^{(a)}_{m_a,k} = 1$, while $P^{(a)}_{m_a,k} = 0$ when $U^{(a)}_{m_a,k} = 0$. The singleton matrix sets $\mathcal{U}^{(a)}$ and $\mathcal{P}^{(a)}$ include the elements of matrices $\mathbf{U}^{(a)}$ and $\mathbf{P}^{(a)}$, respectively. The set of URA strategies in the other $N-1$ cells is characterized by matrix sets $\mathcal{U}^{-a} \triangleq \dot{\times}_{n \in \mathcal{N}, n \neq a} \mathcal{U}^{(n)}$ and $\mathcal{P}^{-a} \triangleq \dot{\times}_{n \in \mathcal{N}, n \neq a} \mathcal{P}^{(n)}$, where $\dot{\times}$ represents the Cartesian product [69].

Ideally, the local URA in the cell a can be formulated as a global optimization problem. The cell a's local objective in terms of power consumption, $L^{(a)}$, is defined as

$$L^{(a)} \triangleq \sum_{m_a=1}^{M^{(a)}} w^{(a)}_{m_a} \sum_{k=1}^{K} P^{(a)}_{m_a,k}, \tag{5.1}$$

where each $w^{(a)}_{m_a}$ denotes session m_a's weight or priority. In consideration of the intra-cell fairness among inner users, near edge users, and far edge users, we define $w^{(a)}_{m_a}$ as

$$w^{(a)}_{m_a} \triangleq \frac{\sum_{n=1,n \neq a}^{N} H^{(a,n)}_{m_a}}{H^{(a,a)}_{m_a}}, \tag{5.2}$$

where each $H^{(a,n)}_{m_a}$ denotes the propagation gain from session m_a to BS n. Note that the definition of user weights is not limited to (2).

There are several constraints for the minimization of $L^{(a)}$. First, if $U^{(a)}_{m_a,k} = 1$, then $P^{(a)}_{m_a,k}$ should be lower bounded by $\hat{Q}^{(a)}_{m_a,k}$, the minimum power for meeting session m_a's SINR requirement, denoted by $\gamma^{(a)}_{m_a}$. Moreover, $P^{(a)}_{m_a,k}$ should also be upper bounded by $\bar{Q}^{(a)}_{m_a,k}$, the maximum power of session m_a on each subchannel k. But if $U^{(a)}_{m_a,k} = 0$, then $P^{(a)}_{m_a,k} = 0$. The BS a cannot make decisions for the other cells to change $\mathcal{U}^{-a}$, so we say $\mathcal{U}^{-a} \equiv \tilde{\mathcal{U}}^{-a}$, where $\tilde{\mathcal{U}}^{-a}$ is a fixed matrix set. But $\mathcal{P}^{(a)}$ and $\mathcal{P}^{-a}$ may interact with each other due to the change of inter-cell interference and corresponding power control, so both $\mathcal{P}^{(a)}$ and $\mathcal{P}^{-a}$ are considered as variables. For example, when session m_a in the cell a and session $m_{a'}$ in another cell a' share the same subchannel k, i.e., $U_{m_a,k}^{(a)} = U_{m_{a'},k}^{(a')} = 1$, increasing $P_{m_a,k}^{(a)}$ can result in an increase in $P_{m_{a'},k}^{(a')}$ to maintain session $m_{a'}$'s SINR due to the increased interference from session m_a. Hence, such constraints should be satisfied not only in the cell a but also in other cells n, namely

$$U_{m_n,k}^{(n)} \hat{Q}_{m_n,k}^{(n)} \le P_{m_n,k}^{(n)} \le U_{m_n,k}^{(n)} \bar{Q}_{m_n,k}^{(n)} \text{ for } n \in \mathcal{N}; m_n \in \mathcal{M}^{(n)}; k \in \mathcal{K}, \quad (5.3)$$

where $\hat{Q}_{m_n,k}^{(n)} \triangleq \dfrac{\gamma_{m_n}^{(n)} (\sum_{n'=1,n'\neq n}^{N} \sum_{m_{n'}=1}^{M^{(n')}} P_{m_{n'},k}^{(n')} H_{m_{n'}}^{(n',n)} + Z_k)}{H_{m_n}^{(n,n)}}$, in which Z_k denotes the noise power on subchannel k. At each BS n, let $I_k^{(n)} \triangleq \sum_{n'=1,n'\neq n}^{N} \sum_{m_{n'}=1}^{M^{(n')}} P_{m_{n'},k}^{(n')} H_{m_{n'}}^{(n',n)}$ denote the aggregated interference from the sessions in other cells n' on subchannel k, which depends on $\mathcal{U}^{-n}$ and $\mathcal{P}^{-n}$. Second, each session m_n's aggregated uplink capacity (per unit bandwidth) in cell n should meet its corresponding QoS requirement, denoted by $\theta_{m_n}^{(n)}$. We have

$$\sum_{k=1}^{K} \log(1 + \frac{P_{m_n,k}^{(n)} H_{m_n}^{(n,n)}}{I_k^{(n)} + Z_k}) \ge \theta_{m_n}^{(n)} \text{ for } n \in \mathcal{N}; m_n \in \mathcal{M}^{(n)}. \quad (5.4)$$

Third, the BS a should not assign more than one session to any subchannel k at the same time to address the issue of intra-cell interference in the uplink. We have

$$\sum_{m_a=1}^{M^{(a)}} U_{m_a,k}^{(a)} \le 1 \text{ for } k \in \mathcal{K}. \quad (5.5)$$

The URA problem in the cell a towards USFR is defined as follows.

Problem 5

$$
\begin{aligned}
\text{Given:} \quad & \mathcal{U}^{-a}; \\
\text{Find:} \quad & \mathcal{U}^{(a)}, \mathcal{P}^{(a)}, \mathcal{P}^{-a}; \\
\text{Minimize:} \quad & L^{(a)}; \\
\text{Subject to:} \quad & (3), (4), (5).
\end{aligned}
$$

Problem 5 is formulated as a mixed-integer non-linear program (MINLP), which is NP-hard in general [60]. Instead of solving it directly, as in [65], we can decouple the complex URA problem in the cell a into two subproblems: TPC by adapting $\mathcal{P}^{(a)}$ and $\mathcal{P}^{-a}$ given fixed $\mathcal{U}^{(a)}$, and SCA by adapting $\mathcal{U}^{(a)}$ given known $\mathcal{P}^{(a)}$ and $\mathcal{P}^{-a}$ as functions of $\mathcal{U}^{(a)}$. Hence, the URA problem can be reduced to the SCA subproblem, which is a binary integer program (BIP). Later in this chapter, we show that finding the optimal solution to the SCA subproblem requires perfect global knowledge and high computational complexity.

5.1.3 A Game-Theoretic Framework

In view of the distributed nature of CR networks, each cell in the multi-cell system has to conduct local URA independently. Because of possible conflicts in coexisting cells' local optimal strategies, we are interested in answering the following questions: (1) Is such a non-cooperative multi-cell decision-making system stable? (2) If the system is stable, how close is the distributed stable point to the global optimal point? (3) Are there any other benefits of optimizing local objective, i.e., power consumption, in each cell, such as spectral efficiency and coexistence fairness? (4) Can the non-cooperative approach be applied to the cooperative case where the coexisting cells coordinate with each other? Therefore, we choose to use game theory to study the global performance of multi-cell URA.

The self-coexistence of CR networks can be modeled as a non-cooperative game, in which each network cell acts as a player. In the URA game, each cell solves Problem 5 independently. Then, minimizing $L^{(n)}$ is equivalent to optimizing cell n's utility. According to the decoupled SCA and TPC subproblems, we adopt a two-level game-theoretic approach to enable USFR. The URA game can be decoupled into SCA and TPC games. In the two-level framework, each acting cell plays the SCA game on the first level. Given each new strategy taken by any cell in the SCA game, the Nash equilibrium is achieved in the TPC game on the second level, which will be proved later. Hence, solving the local URA problem in any acting cell is equivalent to solving the SCA subproblem once while solving the TPC subproblem multiple times (once for each change of SCA strategy by the cell). The acting cells take turns to solve the URA problem, so the URA game is achieved by successive local searches of SCA and TPC strategies optimizing local objectives. Based on the optimal utility gain, each acting cell is able to know whether this two-level URA strategy is beneficial. As soon as nobody can find an improving strategy, a stabilized coexistence pattern is commonly agreed by all the cells in the URA game.

5.2 Game of Transmit Power Control

In this section, we provide a unique optimal solution to the TPC subproblem, and study multi-cell TPC as a non-cooperative game.

5.2.1 Transmit Power Control Subproblem

The existence of binary variables in $\mathcal{U}^{(a)}$ makes Problem 5 costly to solve. If $\mathcal{U}^{(a)}$ is fixed in advance, the URA problem can be reduced to the TPC subproblem. The optimal TPC strategy set, say $\tilde{\mathcal{P}}^N$, depends on the setting of $\mathcal{U}^{(a)}$, say $\mathcal{U}^{(a)} \equiv \tilde{\mathcal{U}}^{(a)}$. In our two-level framework, $\tilde{\mathcal{U}}^{(a)}$ is a SCA strategy taken to solve the SCA subproblem. Thus, $\tilde{\mathcal{U}}^{(a)}$ and $\tilde{\mathcal{P}}^N$ together should satisfy the constraints of Problem 5. Besides (3), (4), and (5) for $a \in \mathcal{N}$, the setting of $\mathcal{U}^{(a)} \equiv \tilde{\mathcal{U}}^{(a)}$ also needs to satisfy the conditions (6) and (8) below to make $\tilde{\mathcal{P}}^N$ feasible to Problem 5. We have

$$U_{m_n,k}^{(n)} \tilde{Q}_{m_n,k}^{(n)} \le U_{m_n,k}^{(n)} \bar{Q}_{m_n,k}^{(n)} \text{ for } n \in \mathcal{N}; m_n \in \mathcal{M}^{(n)}; k \in \mathcal{K}, \tag{5.6}$$

where $\tilde{Q}_{m_n,k}^{(n)} = \hat{Q}_{m_n,k}^{(n)}\big|_{P_{m_{n'},k}^{(n')} = \tilde{P}_{m_{n'},k}^{(n')}}$ and each $\tilde{P}_{m_n,k}^{(n)}$ is from the solution to the equations

$$P_{m_n,k}^{(n)} \equiv U_{m_n,k}^{(n)} \hat{Q}_{m_n,k}^{(n)} \text{ for } n \in \mathcal{N}; m_n \in \mathcal{M}^{(n)}; k \in \mathcal{K}. \tag{5.7}$$

The values of $\tilde{P}_{m_n,k}^{(n)}$ are determined as long as $\tilde{\mathcal{U}}^{(a)}$ is fixed, and so are the values of $\tilde{Q}_{m_n,k}^{(n)}$. Each $\tilde{Q}_{m_n,k}^{(n)}$ is the minimum possible value of corresponding $\hat{Q}_{m_n,k}^{(n)}$ under $\mathcal{U}^{(a)} \equiv \tilde{\mathcal{U}}^{(a)}$. And

$$\sum_{k=1}^{K} \log(1 + U_{m_n,k}^{(n)} \gamma_{m_n}^{(n)}) \ge \theta_{m_n}^{(n)} \text{ for } n \in \mathcal{N}; m_n \in \mathcal{M}^{(n)}. \tag{5.8}$$

Hence, the TPC subproblem in the cell a is defined as follows.

Problem 6

$$\begin{aligned}
\text{Given:} \quad & \mathcal{U}^{(a)}, \mathcal{U}^{-a}; \\
\text{Find:} \quad & \mathcal{P}^{(a)}, \mathcal{P}^{-a}; \\
\text{Minimize:} \quad & L^{(a)}; \\
\text{Subject to:} \quad & (3), (4), (5), (6), (8).
\end{aligned}$$

Lemma 2 *Given that the previously fixed SCA strategy set $\tilde{\mathcal{U}}^N = \tilde{\mathcal{U}}^{(a)} \dot{\times} \tilde{\mathcal{U}}^{-a}$ satisfies (5) for $a \in \mathcal{N}$, (6), and (8), the unique optimal solution of TPC strategy set, say $\tilde{\mathcal{P}}^N = \tilde{\mathcal{P}}^{(a)} \dot{\times} \tilde{\mathcal{P}}^{-a}$, to Problem 6 satisfies (7).*

Proof According to (1) and (2), we know that

$$\frac{\partial L^{(a)}}{\partial P_{m_a,k}^{(a)}} = w_{m_a}^{(a)} > 0 \text{ for } m_a \in \mathcal{M}^{(a)}; k \in \mathcal{K}. \tag{5.9}$$

Obviously, in order to minimize $L^{(a)}$, each $P_{m_a,k}^{(a)}$ in $\mathcal{P}^{(a)}$ needs to be as small as possible. We can see that any inequality relationship $P_{m_a,k}^{(a)} \geq U_{m_a,k}^{(a)} \hat{Q}_{m_a,k}^{(a)}$ in (3) can be rewritten as

$$P_{m_a,k}^{(a)} \geq \sum_{n=1,n\neq a}^{N} \sum_{m_n=1}^{M^{(n)}} c_{m_n,m_a,k}^{(n,a)} P_{m_n,k}^{(n)} + c_{m_a,k}^{(a)} \text{ for } m_a \in \mathcal{M}^{(a)}; k \in \mathcal{K}, \qquad (5.10)$$

in which all the coefficients c's are non-negative. Furthermore, for each $P_{m_n,k}^{(n)}$ on the right-hand side (RHS) of (10), we have

$$P_{m_n,k}^{(n)} \geq \sum_{n'=1,n'\neq n}^{N} \sum_{m_{n'}=1}^{M^{(n')}} c_{m_{n'},m_n,k}^{(n',n)} P_{m_{n'},k}^{(n')} + c_{m_n,k}^{(n)} \text{ for } n \in \mathcal{N}, n \neq a; m_n \in \mathcal{M}^{(n)}; k \in \mathcal{K},$$

$$(5.11)$$

in which all the coefficients c's are non-negative. We can see that any $P_{m_n,k}^{(n)}$ in $\mathcal{P}^{(n)}$ is lower bounded by a linear combination of $P_{m_{n'},k}^{(n')}$ in $\mathcal{P}^{-n}$ with all non-negative coefficients. All the inequality relationships in (10) and (11) are in a cycle, since each $P_{m_a,k}^{(a)}$ on the left-hand side (LHS) of (10) appears on the RHS of (11) as well. Clearly, when the equalities hold for all in (10) and (11), each $P_{m_n,k}^{(n)}$ in $\mathcal{P}^N = \mathcal{P}^{(a)} \dot{\times} \mathcal{P}^{-a}$ reaches its lower bound $U_{m_n,k}^{(n)} \hat{Q}_{m_n,k}^{(n)} = U_{m_n,k}^{(n)} \tilde{Q}_{m_n,k}^{(n)}$. At the same time, the weighted sum $L^{(a)}$ is minimized by $\tilde{\mathcal{P}}^N$ that solves (7) without considering any other constraints. Note that (7) is a system of linear equations, thus $\tilde{\mathcal{P}}^N$ is unique.

We next verify the feasibility of $\tilde{\mathcal{P}}^N$, which is the unique solution that minimizes $L^{(a)}$. It is easy to see that (3) holds given (6) and (7). If (7) holds, the LHS's of (4) and (8) are same, so (4) holds given (7) and (8). And (5) for $a \in \mathcal{N}$ is known to hold. Hence, $\tilde{\mathcal{P}}^N$ satisfies the constraints of Problem 5, and is thus the unique optimal solution to Problem 6. $\qquad\square$

Besides minimizing power consumption as defined by $L^{(a)}$, we are also interested in maximizing the sessions' power efficiency, i.e., uplink capacity per unit power. Again in the form of weighted sum, the cell a's local objective in terms of power efficiency, $E^{(a)}$, is defined as

$$E^{(a)} \triangleq \sum_{m_a=1}^{M^{(a)}} v_{m_a}^{(a)} \sum_{\substack{k=1 \\ U_{m_a,k}^{(a)} \neq 0}}^{K} \frac{\log\left(1 + \frac{P_{m_a,k}^{(a)} H_{m_a}^{(a,a)}}{I_k^{(a)} + Z_k}\right)}{P_{m_a,k}^{(a)}}, \qquad (5.12)$$

where each $v_{m_a}^{(a)}$ can be any positive weight. At the BS a, let $S_{m_a,k}^{(a)} \triangleq \frac{P_{m_a,k}^{(a)} H_{m_a}^{(a,a)}}{I_k^{(a)} + Z_k}$ denote the SINR of session m_a on subchannel k.

Lemma 3 *The unique solution of TPC strategy set that satisfies* (7), *i.e.,* $\tilde{\mathcal{P}}^N$ $= \tilde{\mathcal{P}}^{(a)} \dot{\times} \tilde{\mathcal{P}}^{-a}$, *maximizes* $E^{(a)}$, *subject to the constraints of* Problem 6.

Proof We start from proving that $\tilde{\mathcal{P}}^N$ maximizes

$$\hat{E}^{(a)} \triangleq \sum_{m_a=1}^{M^{(a)}} v_{m_a}^{(a)} \sum_{\substack{k=1 \\ U_{m_a,k}^{(a)} \neq 0}}^{K} \frac{\log(1 + S_{m_a,k}^{(a)})}{S_{m_a,k}^{(a)}}. \tag{5.13}$$

Define $f(s) = \frac{s}{1+s} - \log(1 + s)$. It is trivial to prove that $f(s)$ is a non-increasing function of s. Hence, $f(S_{m_a,k}^{(a)}) \leq f(0) \equiv 0$, since each $S_{m_a,k}^{(a)} \geq 0$ always holds. Then, we have

$$\frac{\partial \hat{E}^{(a)}}{\partial S_{m_a,k}^{(a)}} = v_{m_a}^{(a)} \frac{\frac{S_{m_a,k}^{(a)}}{1+S_{m_a,k}^{(a)}} - \log(1 + S_{m_a,k}^{(a)})}{(S_{m_a,k}^{(a)})^2} \leq 0 \text{ for } m_a \in \mathcal{M}^{(a)}; k \in \mathcal{K}. \tag{5.14}$$

As in the proof for Lemma 2, each $S_{m_a,k}^{(a)}$ is minimized to $U_{m_a,k}^{(a)} \gamma_{m_a}^{(a)}$ by $\tilde{\mathcal{P}}^N$. And $\hat{E}^{(a)}$ is maximized accordingly. To find the relationship between $E^{(a)}$ and $\hat{E}^{(a)}$, we can see that

$$\frac{\log(1 + S_{m_a,k}^{(a)})}{P_{m_a,k}^{(a)}} = \frac{\log(1 + S_{m_a,k}^{(a)})}{S_{m_a,k}^{(a)}} \frac{H_{m_a}^{(a,a)}}{I_k^{(a)} + Z_k} \text{ for } m_a \in \mathcal{M}^{(a)}; k \in \mathcal{K}. \tag{5.15}$$

On the RHS of (15), when each $\frac{\log(1+S_{m_a,k}^{(a)})}{S_{m_a,k}^{(a)}}$ is maximized by $\tilde{\mathcal{P}}^N$, the corresponding $I_k^{(a)}$ is minimized simultaneously. Hence, each $\frac{\log(1+S_{m_a,k}^{(a)})}{P_{m_a,k}^{(a)}}$ is maximized by $\tilde{\mathcal{P}}^N$ that solves (7), and so is the weighted sum $E^{(a)}$ with all positive weights. $\qquad\square$

5.2.2 Multi-cell Transmit Power Control as Non-cooperative Game

To study how each cell behaves in presence of other coexisting cells, we frame multi-cell TPC as a non-cooperative game, which serves as the second level of the URA game as in our two-level framework. In the TPC game, each cell solves Problem 6 independently.

Theorem 7 *The unique solution of TPC strategy set that satisfies* (7), *i.e.,* $\tilde{\mathcal{P}}^N$, *establishes the Nash equilibrium in the non-cooperative TPC game defined by* Problem 6.

Proof According to the fixed point theorem [69] in game theory, two conditions must be satisfied for the existence of Nash equilibrium in the game: a) the strategy space, $\mathbb{P}^N = \dot{\times}_{n \in \mathcal{N}} \mathbb{P}^{(n)}$, for searching $\mathcal{P}^N$ should be a non-empty, compact, and convex subset of certain Euclidean space; b) the utility functions, $L^{(n)}$ for $n \in \mathcal{N}$, should be continuous in $\mathcal{P}^N$ and quasi-convex in $\mathcal{P}^{(n)}$. Due to the bounds defined in (3) for each $P^{(n)}_{m_n,k}$ in $\mathcal{P}^{(n)}$, $\mathbb{P}^{(n)}$ in each cell n is closed and bounded, and thus compact. It is also trivial to show that $\mathbb{P}^{(n)}$ is convex. We know that the Cartesian product of compact and convex sets is still compact and convex. Hence, the first condition is satisfied. As defined in (1), each $L^{(n)}$ is a linear combination of $P^{(n)}_{m_n,k}$ in $\mathcal{P}^{(n)}$ with all positive coefficients. Clearly, $L^{(n)}$ is continuous in $\mathcal{P}^N$. We know that a linear combination of convex functions with positive coefficients is again convex, and every convex function is also quasi-convex. Then, $L^{(n)}$ is quasi-convex in $\mathcal{P}^{(n)}$. Hence, the second condition is satisfied too.

As in the proof for Lemma 2, not only $P^{(a)}_{m_a,k}$ in $\mathcal{P}^{(a)}$ but also $P^{(n)}_{m_n,k}$ in $\mathcal{P}^{-a}$ are minimized to $U^{(n)}_{m_n,k} \tilde{Q}^{(n)}_{m_n,k}$ by $\tilde{\mathcal{P}}^N$. These lower bounds are determined as soon as $\tilde{\mathcal{U}}^N$ is fixed. Regardless the different positive weights in coexisting cells' local objectives, the optimal TPC strategies applied in different cells are always the same unique one, i.e., $\tilde{\mathcal{P}}^N = \tilde{\mathcal{P}}^{(1)} \dot{\times} \tilde{\mathcal{P}}^{-1} = \cdots = \tilde{\mathcal{P}}^{(N)} \dot{\times} \tilde{\mathcal{P}}^{-N}$, that solves (7). According to the fixed point theorem, any fixed point of the best strategy space, $\mathbb{R}^N = \dot{\times}_{n \in \mathcal{N}} \mathbb{R}^{(n)}$, is the Nash equilibrium. Surely, the unique common best strategy set $\tilde{\mathcal{P}}^N$ is a fixed point of $\mathbb{R}^N$ in the TPC game, since any cell's best strategy is always to stay once $\tilde{\mathcal{P}}^N$ has been reached. As a result, the Nash equilibrium is established by $\tilde{\mathcal{P}}^N$ in the TPC game.   □

Theorem 7 offers a guideline of designing TPC algorithms in a multi-cell system. In fact, there are already various existing TPC algorithms that do not rely on perfect inter-cell coordination, e.g., iterative water-filling algorithm [65].

5.3 Game of Subchannel Allocation

In this section, we focus on the SCA subproblem, and try to study multi-cell SCA as a non-cooperative game as well. However, unlike the case of TPC, several undesired facts for the SCA subproblem and game are briefly discussed before going any further.

5.3.1 Subchannel Allocation Subproblem

If $\mathcal{P}^{(a)}$ and $\mathcal{P}^{-a}$ in Problem 5 are treated as known constants when selecting $\mathcal{U}^{(a)}$, the URA problem can be reduced to the SCA subproblem. For each selected $\mathcal{U}^{(a)}$ and fixed $\mathcal{U}^{-a}$, the setting of $\mathcal{P}^{(a)}$ and $\mathcal{P}^{-a}$ can be $\mathcal{P}^N = \mathcal{P}^{(a)} \dot{\times} \mathcal{P}^{-a} \equiv \tilde{\mathcal{P}}^N$, taking advantage of the unique optimal solution to Problem 6 as in Lemmas 2 and 3. Note that $\mathcal{P}^{(a)}$ and $\mathcal{P}^{-a}$ do not keep the same values every time when $\mathcal{U}^{(a)}$ is changed.

Besides (7) for such setting, the conditions (5) for $a \in \mathcal{N}$, (6), and (8) for the setting of $\mathcal{U}^{(a)} \equiv \tilde{\mathcal{U}}^{(a)}$ in Problem 6 also become the constraints of the SCA subproblem in the cell a, which is defined as follows.

Problem 7

$$
\begin{aligned}
\text{Given:} \quad & \mathcal{U}^{-a}; \\
\text{Find:} \quad & \mathcal{U}^{(a)}; \\
\text{Minimize:} \quad & L^{(a)}; \\
\text{Subject to:} \quad & (5), (6), (7), (8).
\end{aligned}
$$

In fact, solving Problem 5 is equivalent to solving Problem 7 once while solving Problem 6 multiple times (once for each change of SCA strategy when solving Problem 7). In our two-level framework, given each SCA strategy, say $\mathcal{U}^{(a)} \equiv \tilde{\mathcal{U}}^{(a)}$, taken by the cell a to solve Problem 7, Problem 5 is solved once to give $\tilde{\mathcal{P}}^N$ under $\mathcal{U}^{(a)} \equiv \tilde{\mathcal{U}}^{(a)}$. Then, the objective $L^{(a)}$ for this SCA strategy $\mathcal{U}^{(a)} \equiv \tilde{\mathcal{U}}^{(a)}$ can be computed under $\mathcal{P}^N \equiv \tilde{\mathcal{P}}^N$. Hence, $L^{(a)}$ can be regarded as a function of only $\mathcal{U}^{(a)}$ when always taking the optimal TPC strategy given each $\mathcal{U}^{(a)}$. Then, Problem 5 is reduced to Problem 7, which is a BIP.

5.3.2 Multi-cell SCA as Non-cooperative Game

Similar to the TPC game, we try to frame multi-cell SCA as a non-cooperative game, in which each cell solves Problem 7 independently. However, the SCA game, which serves as the first level of the URA game, does not have the good properties as the TPC game does.

Corollary 1 *Solving* Problem 7 *for the optimal SCA strategy necessarily requires the perfect global knowledge of $\tilde{\mathcal{U}}^{-a}$, $\tilde{\mathcal{P}}^{-a}$, and $H_{m_n}^{(n,n')}$ for $n \in \mathcal{N}, n \neq a$ in the cell a.*

Proof Obviously, real-time inter-cell coordination is necessary to keep track of $\tilde{\mathcal{U}}^{-a}$ and $\tilde{\mathcal{P}}^{-a}$. Although it is possible to estimate the values of $H_{m_a}^{(a,n')}$ in the cell a by decoding downlink pilot signals from BS's n' [86, 108], that of $H_{m_n}^{(n,n')}$ for $n \neq a$ are hard to get. $\qquad\square$

Corollary 2 *The complexity of searching for the optimal solution to* Problem 7 *is $O(K!)$.*

Proof Due to the setting of $\mathcal{P}^N \equiv \tilde{\mathcal{P}}^N$, any session m_a's uplink capacity on each taken subchannel is always $\log(1 + \gamma_{m_a}^{(a)})$ given $\gamma_{m_a}^{(a)}$. To minimize $L^{(a)}$ and make (8) hold, the number of subchannels taken by each session m_a is $T_{m_a}^{(a)} \triangleq \lceil \frac{\theta_{m_a}^{(a)}}{\log(1+\gamma_{m_a}^{(a)})} \rceil$.

Then, the number of subchannels taken by the cell a is $T^{(a)} = \sum_{m_a=1}^{M^{(a)}} T_{m_a}^{(a)}$. In view of (5), the number of possible solutions for $\mathcal{U}^{(a)}$ is $\frac{K!}{(K-T^{(a)})!}$. Hence, the complexity of search via brute force is on the order of $O(K!)$. $\square$

Corollary 3 *The Nash equilibrium does not always exist in the non-cooperative SCA game defined by* Problem 7.

Proof The discrete strategy space for searching $\mathcal{U}^{(a)}$ cannot satisfy the conditions for the existence of Nash equilibrium as in the proof for Theorem 7. $\square$

5.4 A Two-Level Game-Theoretic Approach

Towards the globally power-efficient and locally fair USFR, we propose a two-level game-theoretic approach. Instead of getting the optimal solution relying on perfect global knowledge and high computational complexity, we resort to a heuristic yet practical approach.

5.4.1 Local Uplink Resource Allocation Algorithm

In the cell a, solving Problem 7 is to find $\tilde{\mathcal{U}}^{(a)}$ that minimizes the weighted sum of sessions' power consumption. According to the user weights defined in (2), the numerator $\sum_{n=1,n\neq a}^{N} H_{m_a}^{(a,n)}$ characterizes session m_a's potential of interference *to* other cells, while the denominator $H_{m_a}^{(a,a)}$ characterizes session m_a's tolerance for interference *from* other cells. The sessions with larger weights are either more likely to cause interference or are more vulnerable to interference. Thus, such weights representing the sessions' priorities can be used to heuristically perform local SCA. Intuitively, near edge users tend to claim dedicated subchannels and need to be scheduled first, while inner users are the most coexistence-friendly ones and can be scheduled at last.

Now, we focus on a certain session, say m_a, that has been scheduled to take subchannels in the cell a. Suppose that the perfect knowledge of $\tilde{\mathcal{U}}^{-a}$, $\tilde{\mathcal{P}}^{-a}$, and $H_{m_n}^{(n,a)}$ for $n \neq a$ is not available. The only information that can be locally obtained by the BS a is $I_k^{(a)}$, i.e., the aggregated interference measurement on each subchannel k. Hence, the local SCA decisions for the session m_a is merely based on the interference measurements $\hat{I}_k^{(a)} \triangleq I_k^{(a)}(\tilde{\mathcal{P}}^N|_{U_{m_a,k}^{(a)}=0})$ or $\bar{I}_k^{(a)} \triangleq I_k^{(a)}(\tilde{\mathcal{P}}^N|_{U_{m_a,k}^{(a)}=1})$, where $\hat{I}_k^{(a)}$ ($\bar{I}_k^{(a)}$) is measured when no session (the session m_a) takes subchannel k and $\tilde{\mathcal{P}}^N$ is generated in the TPC game under the fixed $\tilde{\mathcal{U}}^{-a}$ and $U_{m_a,k}^{(a)} = 0$ ($U_{m_a,k}^{(a)} = 1$). Intuitively, the session m_a is supposed to take $T_{m_a}^{(a)}$ subchannels with the lowest $\bar{I}_k^{(a)}$ to meet the required SINR and QoS. However, the values of $\bar{I}_k^{(a)}$ are costly to get if

---

Algorithm 5 Local URA algorithm

1: **sort** the sequence of sessions m_a for $m_a \in \mathcal{M}^{(a)}$ by $w_{m_a}^{(a)}$ in descending order, and store the sorted sequence in array $\mathbf{M}(i)$ for $i = 1 \to M^{(a)}$

2: **set** $\mathcal{U}^{(a)} = \{\mathbf{0}\}$, and measure $\hat{I}_k^{(a)}$ for $k \in \mathcal{K}$ generated in the TPC game

3: **sort** the sequence of subchannel k for $k \in \mathcal{K}$ by $\hat{I}_k^{(a)}$ in ascending order, and store the sorted sequence in array $\mathbf{K}(j)$ for $j = 1 \to K$

4: **for** $i = 1 \to M^{(a)}$ **do**

5: **set** $m_a = \mathbf{M}(i)$

6: **for** $j = 1 \to K$ **do**

7:   **set** $k = \mathbf{K}(j)$

8:   **if** $\sum_{m=1}^{M^{(a)}} U_{m,k}^{(a)} == 0$ **then**

9:    **set** $U_{m_a,k}^{(a)} = 1, \cdots, U_{m_a,k+T_{m_a}^{(a)}-1}^{(a)} = 1$

10:    **break**

11:   **end if**

12: **end for**

13: **end for**

14: **measure** $\bar{I}_k^{(a)}$ for k whose $\sum_{m=1}^{M^{(a)}} U_{m,k}^{(a)} == 1$ generated in the TPC game, and adjust $P_{m_a,k}^{(a)}$ to meet the required SINR and QoS if $U_{m_a,k}^{(a)} == 1$

---

inter-cell coordination is not assumed, since the session m_a has to try each subchannel k by triggering the TPC game to create $\bar{I}_k^{(a)}$. Instead, the session m_a can take $T_{m_a}^{(a)}$ subchannels with the lowest $\hat{I}_k^{(a)}$ thanks to the connection between $\hat{I}_k^{(a)}$ and $\bar{I}_k^{(a)}$. In most cases, $\bar{I}_{k_1}^{(a)} < \bar{I}_{k_2}^{(a)}$ is true if $\hat{I}_{k_1}^{(a)} < \hat{I}_{k_2}^{(a)}$ for the same session m_a. Therefore, a heuristic solution of SCA strategy to Problem 7 can be given by Algorithm 5.

Corollary 4 *The complexity of running* Algorithm 5 *is* $O(K M^{(a)})$.

Proof The major cost for running Algorithm 5 comes from the two sorting operations in lines 1 and 3 and the nested FOR loops between lines 4 and 13. Sorting sessions leads to the complexity of $O(M^{(a)} \log M^{(a)})$ via e.g., merge sort. Likewise, sorting subchannels incurs $O(K \log K)$. The nested FOR loops result in $O(K M^{(a)})$. Because $K > M^{(a)}$ and possibly $M^{(a)} > \log K$, the complexity of running Algorithm 5 is on the order of $O(K M^{(a)})$. $\qquad\square$

5.4.2 Two-Level Game-Theoretic Algorithm

Due to the fact that the SCA game does not always converge, a cost function [69] should be considered in the definition of utility. We start from a cooperative URA game where coexisting cells share the same global objective, G, which can be defined as

$$G \triangleq \sum_{n=1}^{N} L^{(n)} = \sum_{n=1}^{N} \sum_{m_n=1}^{M^{(n)}} w_{m_n}^{(n)} \sum_{k=1}^{K} P_{m_n,k}^{(n)}, \tag{5.16}$$

which is the weighted sum of sessions' power consumption in all cells. It is trivial to show that the cooperative game in which each cell solves Problem 5 with common global objective G can converge to the globally optimal coexistence pattern with unique optimal value of G. We can rewrite (16) as $G = L^{(a)} + C_g^{(a)}$, where the cell a's global cost function can be rewritten as

$$C_g^{(a)} \triangleq \sum_{n=1,n\neq a}^{N} \sum_{m_n=1}^{M^{(n)}} \sum_{n'=1,n'\neq n}^{N} \sum_{k=1}^{K} \frac{P_{m_n,k}^{(n)} H_{m_n}^{(n,n')}}{H_{m_n}^{(n,n)}}. \tag{5.17}$$

On the RHS of (17), each $P_{m_n,k}^{(n)} H_{m_n}^{(n,n')}$ represents the interference from session m_n to BS n'. In the cell a, however, only the sum of $P_{m_n,k}^{(n)} H_{m_n}^{(n,a)}$ on each subchannel k, i.e., $I_k^{(a)}$, is known. Hence, the cell a's local cost function, denoted by $C_l^{(a)}$, can be defined as

$$C_l^{(a)} \triangleq \delta^{(a)} \sum_{k=1}^{K} I_k^{(a)}, \tag{5.18}$$

where $\delta^{(a)}$ denotes a positive price factor. Note that unlike the commonly defined cost functions, such as $\sum_{m_a=1}^{M^{(a)}} \sum_{k=1}^{K} \log(P_{m_a,k}^{(a)})$ [69], that change in the opposite direction of the change in $L^{(a)}$ while adapting any $P_{m_a,k}^{(a)}$, the defined $C_l^{(a)}$ and $L^{(a)}$ can be reduced together most of the time. When the cell a performs local URA to minimize $L^{(a)}$, most of the coexisting cells in the multi-cell system can reduce power simultaneously for the drop of interference from the cell a.

After replacing $L^{(a)}$ with $L^{(a)} + C_l^{(a)}$ in Problem 6 and 7, we get the revised TPC and SCA subproblems in the cell a, respectively.

Lemma 4 *Given that the previously fixed SCA strategy set $\tilde{\mathcal{U}}^N = \tilde{\mathcal{U}}^{(a)} \dot\times \tilde{\mathcal{U}}^{-a}$ satisfies (5) for $a \in \mathcal{N}$, (6), and (8), the unique optimal solution of TPC strategy set, say $\tilde{\mathcal{P}}^N = \tilde{\mathcal{P}}^{(a)} \dot\times \tilde{\mathcal{P}}^{-a}$, to Problem 4 satisfies (7). Globally, $\tilde{\mathcal{P}}^N$ further establishes the Nash equilibrium in the non-cooperative TPC game defined by Problem 4.*

Proof The proof follows the same logic as that for Lemma 2 and Theorem 7.   □

As above, solving the revised Problem 1 with objective $L^{(a)} + C_l^{(a)}$ is equivalent to solving Problem 5 once while solving Problem 4 multiple times (once for each change of SCA strategy when solving Problem 5). To guarantee the convergence of the URA game to a commonly agreed coexistence pattern, the adaptation of $\delta^{(n)}$ in each $C_l^{(n)}$ is necessary to eliminate the interest conflicts among greedy cells in the game. We do not assume available perfect global knowledge and centralized decision maker like virtual referee [40]. Intuitively, it is difficult to ensure that the non-cooperative URA

Algorithm 6 Two-level game-theoretic algorithm

1: **set** $\delta^{(a)} = \delta_0^{(a)}$ ($\delta_0^{(a)}$ is a positive constant), and begin to participate in the SCA
   game
2: **loop**
3:    **set** the backoff timer for a random time interval
4:    **repeat**
5:      **do** the countdown of backoff timer
6:    **until** the backoff timer expires
7:    **measure** $\tilde{I}_k^{(a)}$ for $k \in \mathcal{K}$, and compute $\tilde{L}^{(a)}$ under the current $\mathcal{U}^N$ and $\tilde{\mathcal{P}}^N$
     generated in the TPC game
8:    **set** $L_0^{(a)} = \tilde{L}^{(a)}$, $I_0^{(a)} = \sum_{k=1}^K \tilde{I}_k^{(a)}$, and $F_0^{(a)} = L_0^{(a)} + \delta_0^{(a)} I_0^{(a)}$, and record
     $\mathcal{U}_0^{(a)} = \mathcal{U}^{(a)}$
9:    **run** Algorithm 5
10:    **set** $L^{(a)} = \tilde{L}^{(a)}$, $I^{(a)} = \sum_{k=1}^K \tilde{I}_k^{(a)}$, and $F^{(a)} = L^{(a)} + \delta^{(a)} I^{(a)}$ under the
     updated $\mathcal{U}^N$ and $\tilde{\mathcal{P}}^N$ from Algorithm 5
11:    **if** $F^{(a)} \geq F_0^{(a)}$ **then**
12:      **set** $\mathcal{U}^{(a)} = \mathcal{U}_0^{(a)}$ (give up acting in the SCA game)
13:    **end if**
14:    **if** !($L^{(a)} \leq L_0^{(a)}$ && $I^{(a)} \leq I_0^{(a)}$) **then**
15:      **update** $\delta^{(a)} = \delta^{(a)} + \Delta^{(a)}$ ($\Delta^{(a)}$ is a positive constant)
16:    **end if**
17: **end loop**

---

game defined by $C_l^{(a)}$ always converges to the so-called globally optimal coexistence pattern created by the cooperative URA game defined by $C_g^{(a)}$. Moreover, even such an optimal coexistence pattern can be established via perfect inter-cell coordination, it can be subject to frequent change due to dynamic spectrum environment and potential user mobility. Combined with Algorithm 5, a heuristic yet distributed Algorithm 6 is proposed. We next prove that the cells running Algorithm 6 independently can converge to a common stable coexistence pattern without inter-cell coordination.

In Algorithm 6, the loop between lines 2 and 17 describes the cell a's behavior in the SCA game. Whenever the cell a takes a new SCA strategy, the TPC game as in lines 2 and 14 in Algorithm 5 is played and achieves the Nash equilibrium as in Lemma 4. The use of backoff timer between lines 3 and 6 aims to simplify the scheduling of acting cells in the multi-cell system. The time interval for the backoff timer can be set in seconds. When the time for solving the URA problem is relatively short, the coexisting cells are expected to take turns to act in the URA game. Then, the non-cooperative URA game is achieved by successive local searches of SCA and TPC strategies optimizing local objectives. The adaptation of $\delta^{(a)}$ in line 15 is to make the cell a contribute to the Nash equilibrium in the URA game, which is proved below.

Theorem 8 *In the non-cooperative URA game defined by* Problems 4 *and* 5, *the cells that run* Algorithm 6 *independently agree on a common stable coexistence pattern under the established Nash equilibrium without inter-cell coordination.*

Proof As in Lemma 4, the TPC game defined by Problem 4 can always achieve the Nash equilibrium whenever any cell in the game changes its SCA strategy. Then, we verify that the cells running Algorithm 6 can agree on a common $\tilde{\mathcal{U}}_0^N = \times_{n \in \mathcal{N}} \tilde{\mathcal{U}}_0^{(n)}$ in the SCA game, and nobody has the intention of changing the converged $\tilde{\mathcal{U}}_0^N$.

According to the price adaptation strategy in Algorithm 6, $\delta^{(a)}$ in the cell a is increased by $\Delta^{(a)}$ if the taken SCA strategy does not contribute to the desired Nash equilibrium. The cell a contributes to the Nash equilibrium only when $L^{(a)} \leq L_0^{(a)}$ and $I^{(a)} \leq I_0^{(a)}$. Hence, $\delta^{(a)}$ is non-decreasing. The cell a changes subchannel allocation only when $F^{(a)} < F_0^{(a)}$. Hence, $F_0^{(a)}$ is decreasing. If $\delta^{(a)}$ has been increased to always make $F^{(a)} \geq F_0^{(a)}$ hold, the cell a no longer takes any action in the SCA game. Then, the value of $F_0^{(a)}$ keeps unchanged, say $F_0^{(a)} \equiv \tilde{F}_0^{(a)}$. The adaptation of $\delta^{(n)}$ for $n \neq a$ is similar. Once all the cells give up acting in the SCA game, a certain common $\tilde{\mathcal{U}}_0^N$ is agreed. The Nash equilibrium is then established by the TPC game under the fixed $\tilde{\mathcal{U}}_0^N$.

Now we prove that the achieved common coexistence pattern is stable. The Lyapunov's direct stability theorem in nonlinear control theory: if there exists a Lyapunov function on a region containing a critical point, which is set as the origin, then the origin is a stable critical point and all solutions originating in the region approach the origin asymptotically [63]. At the convergence point, $F_0^{(n)} \equiv \tilde{F}_0^{(n)}$ for $n \in \mathcal{N}$. Hence, the converged coexistence pattern is a critical point. We can further show that $\sum_{n=1}^{N}(F_0^{(n)} - \tilde{F}_0^{(n)})$ is a Lyapunov function. If $\sum_{n=1}^{N}(F_0^{(n)} - \tilde{F}_0^{(n)})$ is a Lyapunov function, it needs to satisfy the following three conditions:

1. The summation $\sum_{n=1}^{N}(F_0^{(n)} - \tilde{F}_0^{(n)}) = 0$ at the origin, i.e., the convergence point, and $\sum_{n=1}^{N}(F_0^{(n)} - \tilde{F}_0^{(n)}) > 0$ for other coexistence patterns;
2. The derivative $\frac{d \sum_{n=1}^{N} F_0^{(n)}}{d\mathbf{p}} > 0$, where $\mathbf{p}$ is any vector radiating from the origin;
3. The derivative $\frac{d \sum_{n=1}^{N} F_0^{(n)}}{dt} < 0$, where t is time.

We know that $\sum_{n=1}^{N} F_0^{(n)}$ reaches its lower bound $\sum_{n=1}^{N} \tilde{F}_0^{(n)}$ at the convergence point. Hence, condition (1) is satisfied. The adaptation of $\delta^{(n)}$ for $n \in \mathcal{N}$ can make $\sum_{n=1}^{N} F_0^{(n)}$ decreasing, so the convergence point is the lowest point in terms of $\sum_{n=1}^{N} F_0^{(n)}$. Hence, condition (2) is satisfied. We know that $\sum_{n=1}^{N} F_0^{(n)}$ is decreasing in time. Hence, condition (3) is satisfied. Then, we can say that $\sum_{n=1}^{N}(F_0^{(n)} - \tilde{F}_0^{(n)})$ is a Lyapunov function. Based on the Lyapunov's direct stability theorem, we conclude that the converged common coexistence pattern is stable. $\qquad \square$

The converged coexistence pattern is achieved by successive local searches of SCA and TPC strategies optimizing local objectives. We have shown that the solution to the local SCA subproblem may not be optimal without having perfect global knowledge and requiring high computational complexity. Furthermore, successive local searches do not guarantee the resulting coexistence pattern to be unique and globally optimal. Hence, the proposed two-level game-theoretic approach as in Algorithm 6 is a heuristic. However, our approach does not rely on any inter-cell coordination and has low complexity. It is always based on the actual local interference measurements and does not need any estimates of environment. Later we show that the converged coexistence pattern in the non-cooperative URA game is usually close enough to the globally optimal coexistence pattern obtained in the cooperative URA game.

5.4.3 Implementation Variants

So far, our discussion of USFR was based on the definition of user weights as in (2) and the assumption that coexisting networks belong to different network operators. In fact, however, there are some other implementation variants of USFR that can be considered.

Variant 1 *The first variant of USFR technique is obtained by modifying the definition of $w_{m_a}^{(a)}$ in Algorithm 5, which is not limited to (2) and can be set on demand. The purpose of optimizing the weighted sum of sessions' power consumption or power efficiency is to enhance intra-cell fairness locally as well as spectrum utilization globally.*

Here, we present two alternate definitions for the user weight $w_{m_a}^{(a)}$. The first definition reflects the fact that user terminals in each cell may not be equal in terms of spectrum access priority. For instance, some users may pay a higher price for higher priority access. If such prioritized spectrum access is considered, $w_{m_a}^{(a)}$ can be defined as

$$w_{m_a}^{(a)} \triangleq \alpha p_{m_a}^{(a)} + (1 - \alpha)\bar{w}_{m_a}^{(a)}, \tag{5.19}$$

where $p_{m_a}^{(a)}$ denotes session m_a's predefined normalized priority; $\bar{w}_{m_a}^{(a)}$ denotes session m_a's normalized weight based on (2); and α is a convex combination factor.

The second definition reflects the fact that a BS, say a, may have difficulty in measuring $\sum_{n=1,n\neq a}^{N} H_{m_a}^{(a,n)}$ locally for each session m_a, since the values of $H_{m_a}^{(a,n)}$ for mobile/portable users and far-away BS's may be very difficult to calculate or estimate accurately. Because precise values of $w_{m_a}^{(a)}$ are not necessarily required for uplink scheduling, an approximation of $\sum_{n=1,n\neq a}^{N} H_{m_a}^{(a,n)}$ can be used instead. Then, $w_{m_a}^{(a)}$ in terms of worst-case interference can be defined as

$$w_{m_a}^{(a)} \triangleq \frac{\bar{H}_{m_a}^{(a,n)}}{H_{m_a}^{(a,a)}}, \tag{5.20}$$

where $\bar{H}_{m_a}^{(a,n)} = \max\limits_{n \in \mathcal{N}, n \neq a} \{H_{m_a}^{(a,n)}\}$, or $\bar{H}_{m_a}^{(a,n)}$ is the propagation gain from session m_a to the nearest BS n if propagation gain is inversely proportional to distance. Moreover, if $\bar{H}_{m_a}^{(a,n)} = 1$, the definition is reduced to the SFR's way of distinguishing inner and edge users in the downlink.

Variant 2 *The second variant of USFR technique is applicable to a single-operator coexistence scenario. In our previous problem formulation, we assumed a multi-operator case where each coexisting cell is controlled by a distinct network operator. Here, we assume that a single network operator centrally manages all the N co-channel cells and $M = \sum_{n=1}^{N} M^{(n)}$ uplink sessions. In such a case, in addition to SCA and TPC, user-cell association (UCA) needs to be considered. Each session m for $m \in \mathcal{M} \triangleq \{1, \cdots, M\}$ needs to decide which cell to associate with. Hence, an extra set of binary variables $\mathcal{R}^{(n)} \triangleq \{R_m^{(n)}; m \in \mathcal{M}\}$ is used to characterize the association of sessions in each cell n. Each $R_m^{(n)}$ is equal to 1 (0) when session m associates (does not associate) with cell n.*

In the above single-operator scenario, multi-cell URA can be formulated as a centralized optimization problem with the global objective G defined in (16). Based on (3), (4), and (5), the constraints for this problem are as follows. First, each $P_{m,k}^{(n)}$ should be bounded. We have

$$R_m^{(n)} U_{m,k}^{(n)} \hat{Q}_{m,k}^{(n)} \leq P_{m,k}^{(n)} \leq R_m^{(n)} U_{m,k}^{(n)} \bar{Q}_{m,k}^{(n)} \text{ for } n \in \mathcal{N}; m \in \mathcal{M}; k \in \mathcal{K}. \quad (5.21)$$

Second, each session m's QoS requirement should be met, and hence we have

$$\sum_{n=1}^{N} \sum_{k=1}^{K} \log(1 + \frac{R_m^{(n)} P_{m,k}^{(n)} H_m^{(n,n)}}{I_k^{(n)} + Z_k}) \geq \theta_m^{(n)} \text{ for } m \in \mathcal{M}. \quad (5.22)$$

Third, each subchannel k should be allocated to no more than one session in cell n. We have

$$\sum_{m=1}^{M} R_m^{(n)} U_{m,k}^{(n)} \leq 1 \text{ for } n \in \mathcal{N}; k \in \mathcal{K}. \quad (5.23)$$

Fourth, each session m should only be associated with one cell, and hence we have

$$\sum_{n=1}^{N} R_m^{(n)} = 1 \text{ for } m \in \mathcal{M}. \quad (5.24)$$

Hence, the centralized single-operator optimization problem (SOP) is defined as follows.

Problem 8

$$\text{Find:} \quad \mathcal{R}^{(n)}, \mathcal{U}^{(n)}, \mathcal{P}^{(n)} \text{ for } n \in \mathcal{N};$$
$$\text{Minimize:} \quad G;$$
$$\text{Subject to:} \quad (21), (22), (23), (24).$$

Solving Problem 8 needs to determine $\mathcal{R}^{(n)}$. Fortunately, the values of $\mathcal{R}^{(n)}$ can be fixed in advance to simplify Problem 6 as the following corollary prescribes.

Corollary 5 *Each session m always gets associated with the cell $n = \underset{n \in \mathcal{N}}{\arg\max}\{H_m^{(n,n)}\}$ for the minimization of G, in which each $w_m^{(n)}$ is defined in (2).*

Proof To minimize G, each session m's $w_m^{(n)}$ needs to be minimized. When a session m registers to the cell $n = \underset{n \in \mathcal{N}}{\arg\max}\{H_m^{(n,n)}\}$, we have $\underset{n \in \mathcal{N}}{\min} w_m^{(n)}$

$$= \frac{\sum_{n=1}^{N} H_m^{(n,n)} - \underset{n \in \mathcal{N}}{\max}\{H_m^{(n,n)}\}}{\underset{n \in \mathcal{N}}{\max}\{H_m^{(n,n)}\}}. \qquad \square$$

Therefore, whenever a session becomes active, it first detects downlink signals from nearby BS's and then associates with the cell n with $\underset{n \in \mathcal{N}}{\max}\{H_m^{(n,n)}\}$ or simply the nearest cell. After the values of $\mathcal{R}^{(n)}$ are fixed, we can solve Problem 6 by the previously defined cooperative URA game, where our approach is still applicable. Note that USFR is still decentralized in the single-operator case. Although centralized optimization can be conducted by the operator, our approach does not require dedicated decision-making infrastructure and high computational complexity.

5.5 Simulation Results

In this section, real-time performance of USFR achieved by Algorithm 6 is evaluated. We focus on a self-coexistence scenario, as in Fig. 5.2, where $N = 7$ CR network cells are forced to share a single TV white-space channel. Each cell is centrally controlled by a BS, and direct inter-cell coordination is not needed. In each cell n, $M^{(n)}$ uplink sessions are generated and randomly distributed. Through simulations, we evaluate a number of different definitions for the user weight: D0 as defined in (2), D1 as defined in (19), D2 as defined in (20) with $\bar{H}_{m_a}^{(a,n)} = \underset{n \in \mathcal{N}, n \neq a}{\max}\{H_{m_a}^{(a,n)}\}$, and D3 as defined in (20) with $\bar{H}_{m_a}^{(a,n)} = 1$. We also apply Corollary 5 to each multi-operator (MO) scenario, and study the corresponding single-operator (SO) case. The average cell radius is 4 km, and the propagation gains are computed via log-distance path-loss model with exponent 3.2. The other parameters are set as follows: $\gamma_{m_n}^{(n)} = 30, T_{m_n}^{(n)} = 1, \bar{Q}_{m_n,k}^{(n)} = 100$ mW, Z_k is set by noise density -174 dBm/Hz and TV channel width 6 MHz. The coexistence patterns created by Algorithm 6 under different combinations of total number of sessions $M \triangleq \sum_{n=1}^{7} M^{(n)}$ and degree of

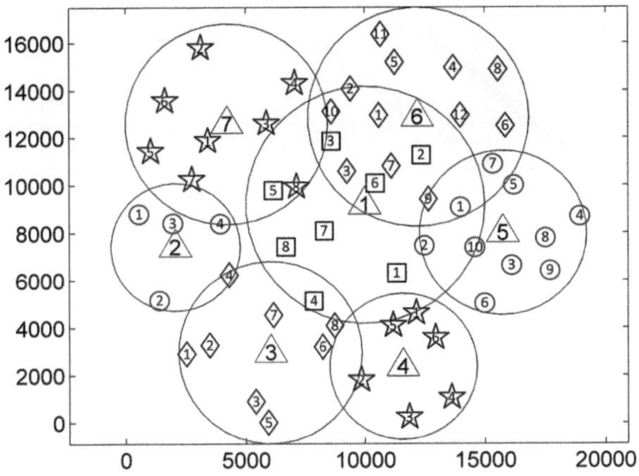

Fig. 5.2 A 7-cell simulation scenario: *Triangles* represent BS's and *other shapes* represent sessions; *Numbers* represent ID numbers of BS's and sessions; X/Y coordinates are in meters

spectrum sharing $\kappa \triangleq \frac{K}{M}$ are analyzed. In the following paragraphs, we discuss the various attributes of USFR based on our simulation experiment results.

First, we evaluate the convergence of USFR. The empirical cumulative distribution function (CDF) of the number of iterations for converging to a common coexistence pattern is shown in Fig. 5.3. Here, one iteration represents a change of URA strategy by any cell in the 7-cell system. The starting point of $\mathcal{U}_0^N$ in Algorithm 6 is any feasible point to Problem 5. We can see that the convergence of USFR is relatively fast even if the shared channel is heavily utilized.

Second, we evaluate the spectrum utilization efficiency of USFR. The distributed USFR is compared with the centralized frequency planning in Fig. 5.4 in terms of the minimum K required to accommodate M sessions. With the global knowledge of the 7-cell topology as in Fig. 5.2, the frequency planning typically requires $\frac{3M}{7}$ subchannels to accommodate M sessions with $T_{m_n}^{(n)} = 1$. This is because the cells 2, 4, and 6 can share the same subchannels safely, and so can the cells 3, 5, and 7. We can see that our USFR under various definitions of user weights generally achieves a similar level of spectrum utilization without inter-cell coordination. In each cell, most inner users can coexist with edge users, probably far edge users, in other cells with acceptable mutual interference. User weight definition D0 outperforms the other definitions due to the use of the greater amount of information on the spectrum environment. Nonetheless, D2 and D3 achieve reasonable spectrum utilization. For the same number of subchannels, from Fig. 5.4, we can also get the maximum number of sessions that can always be supported by the seven cells. If the number of sessions goes beyond the maximum limit, a feasible solution to the URA problem may not be guaranteed. Hence, some sessions have to be blocked if M is too large for a certain K. The blocking strategy depends on the cells' locations and coverage ranges, and the sessions' density, distribution, priority and fairness, etc. Note that the user weights defined in (2) should not be used to block users or sessions, since the

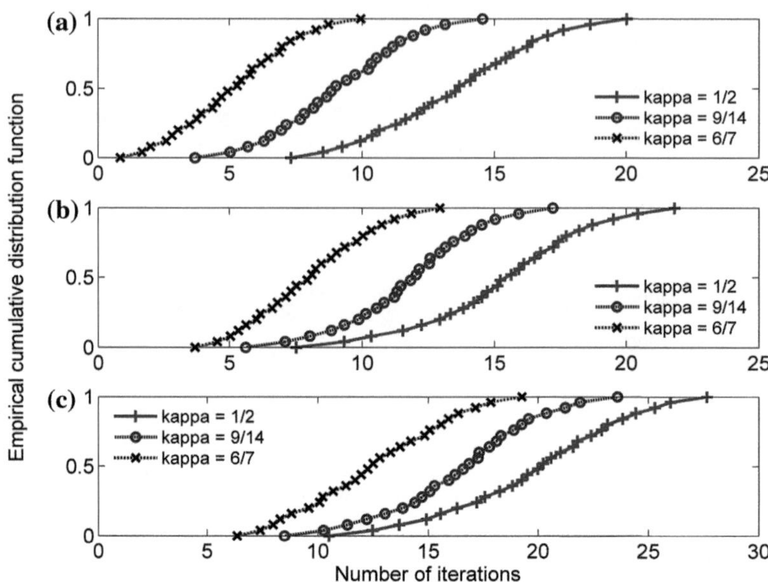

Fig. 5.3 Convergence of USFR ($\Delta = 1 \times 10^{10}$): **a** $M = 28$ (*top*); **b** $M = 56$ (*middle*); **c** $M = 112$ (*bottom*)

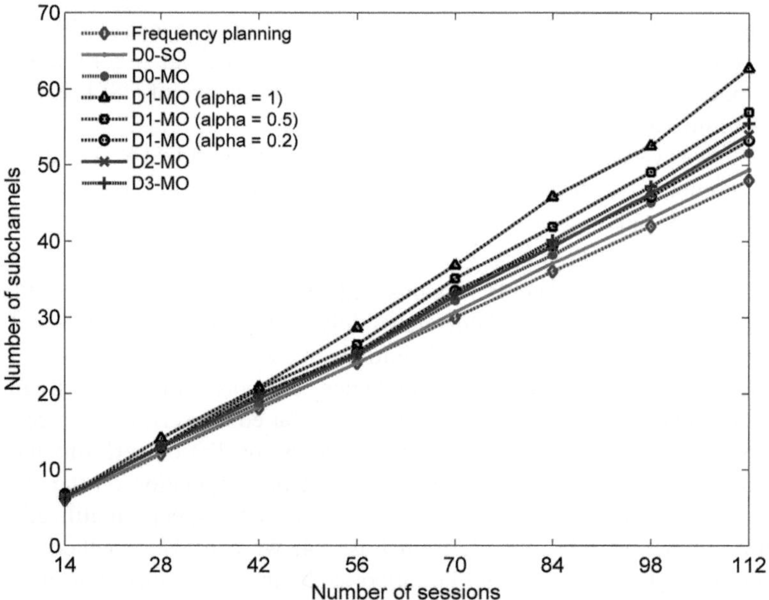

Fig. 5.4 Spectrum utilization of USFR

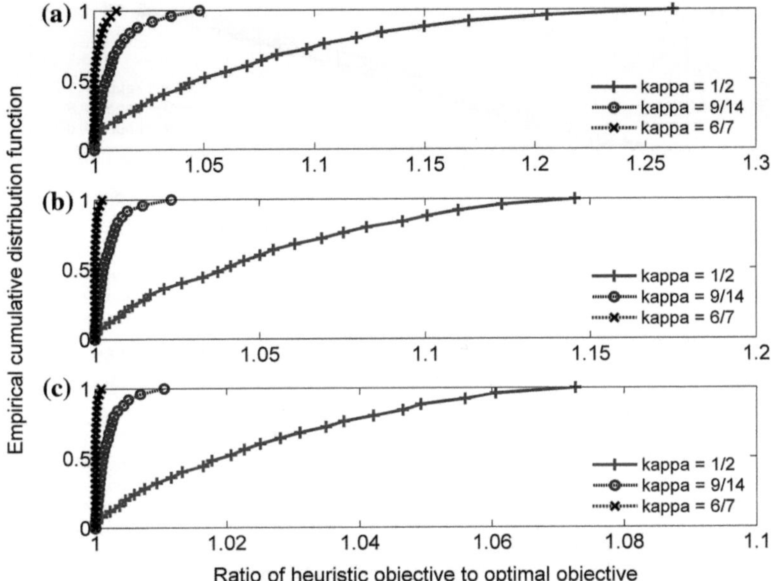

Fig. 5.5 Power consumption of USFR in MO scenarios: **a** $M = 28$ (*top*); **b** $M = 56$ (*middle*); **c** $M = 112$ (*bottom*)

weights only determine the quality of granted subchannels in a feasible solution to the URA problem. A session with a low weight value still has access to low-quality subchannels in a feasible solution. Another type of priority should be used to make an unfeasible solution feasible by blocking users or sessions.

Third, we evaluate the power consumption and power efficiency of USFR. The objective ratio $\frac{heuristic\ \text{value of } G}{optimal\ \text{value of } G}$ is plotted in Fig. 5.5. The heuristic G is computed by using the common coexistence pattern achieved in the non-cooperative URA game under $C_l^{(n)}$, while the optimal G is computed by using the globally optimal coexistence pattern obtained in the cooperative URA game under $C_g^{(n)}$. On the one hand, the heuristic coexistence pattern is generated by successive local searches of SCA and TPC strategies optimizing local objectives, so different orders of cell scheduling may result in different heuristic values of G. In Algorithm 6, we employ random cell scheduling to avoid inter-cell coordination. On the other hand, the optimal coexistence pattern is created by coordinating the cells to optimize the same global objective. The optimal value of G is unique, which cannot be improved by any cell via inter-cell coordination. Both heuristic and optimal coexistence patterns have the same format, which includes $\mathcal{U}^{(n)}$ and $\mathcal{P}^{(n)}$ for $n \in \mathcal{N}$, and G is a function of $\mathcal{U}^{(n)}$ and $\mathcal{P}^{(n)}$, so we can compare the heuristic solutions with the optimal solution. From Fig. 5.5, we can see that our USFR achieves near-optimal performance in most cases, especially when the shared channel is not over saturated. The ratio $\frac{\text{average heuristic value of } G \text{ for SO}}{\text{average heuristic value of } G \text{ for MO}}$ is shown in Fig. 5.6. We can see a significant

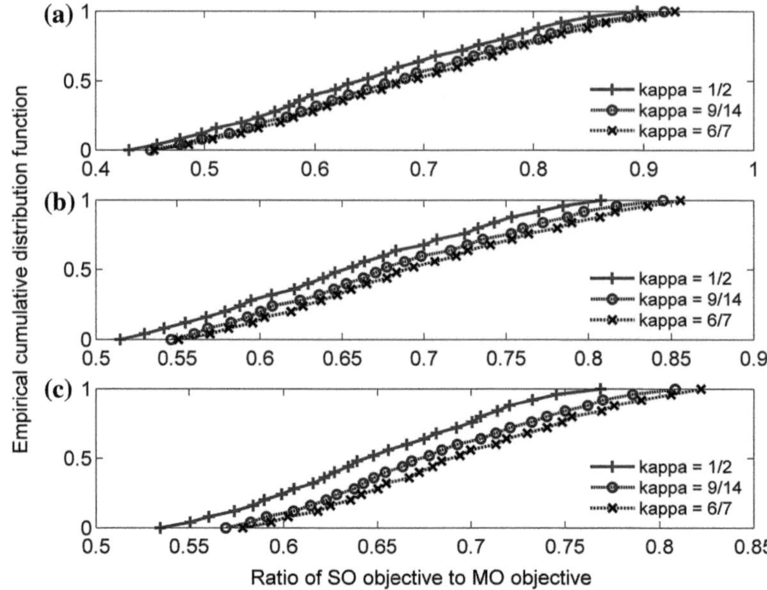

Fig. 5.6 Power consumption of USFR in SO scenarios: **a** $M = 28$ (*top*); **b** $M = 56$ (*middle*); **c** $M = 112$ (*bottom*)

advantage in minimizing G if the seven cells are controlled by one operator. This is due to the reduction of the cells' RF footprints according to Corollary 5. In Fig. 5.7, the ratios $\dfrac{\text{average power efficiency for D1/D2/D3}}{\text{average power efficiency for D0}}$ are compared. For a fair comparison, the power efficiency (without user weights) is defined by the ratio of the sum of sessions' uplink capacity to the sum of sessions' transmit power. We can see that D1 with a small value of α, D2, and D3 achieve similar power efficiency.

Finally, we evaluate the intra-cell fairness of USFR. Firstly, we study a specific scenario as illustrated in Fig. 5.2. The sessions' received power levels at their home BS's are shown in Fig. 5.8, in which each set of bars represents the received power levels at a certain BS and the bars in each set are ordered by sessions' ID numbers and distinguished by different colors. As we can see, the received power levels at the BS are generally the same regardless of the sessions' link distances. The near edge users (that are associated with sessions 2, 3, 4, and 5) in cell 1 tend to utilize subchannels exclusively. However, the inner users (that are associated with sessions 6 and 7) in cell 1 have to coexist with users in other cells, and thus need to receive signals at a higher received power to guarantee the minimum required SINR. Next, we focus on a general case. In each cell n, the fairness factor is defined by $\dfrac{\max_{mn\in\mathcal{M}^{(n)},k\in\mathcal{K},U_{mn,k}^{(n)}\neq 0}\{P_{mn,k}^{(n)}H_{mn}^{(n,n)}\}}{\min_{mn\in\mathcal{M}^{(n)},k\in\mathcal{K},U_{mn,k}^{(n)}\neq 0}\{P_{mn,k}^{(n)}H_{mn}^{(n,n)}\}}$, as shown in Fig. 5.9. We can see that the proposed USFR technique always tries to

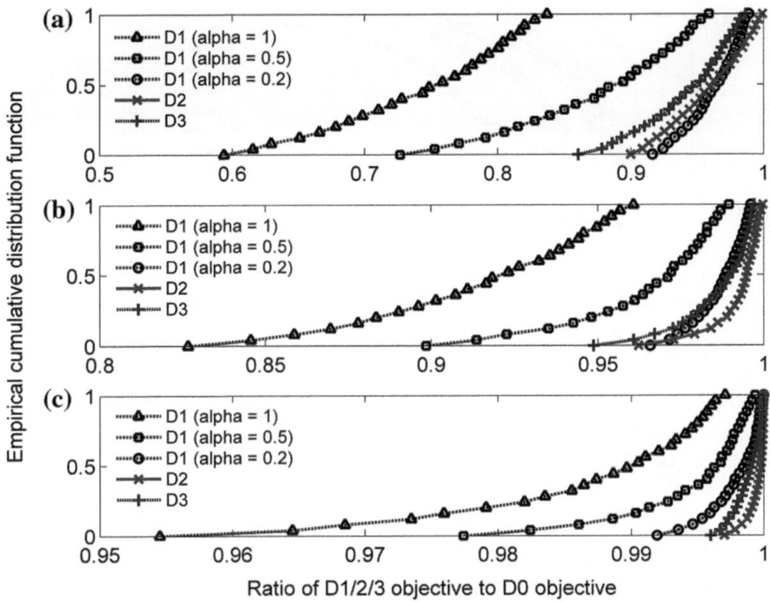

Fig. 5.7 Power efficiency of USFR under alternate definitions of user weights ($M = 56$): **a** $\kappa = 1/2$ (*top*); **b** $\kappa = 9/14$ (*middle*); **c** $\kappa = 6/7$ (*bottom*)

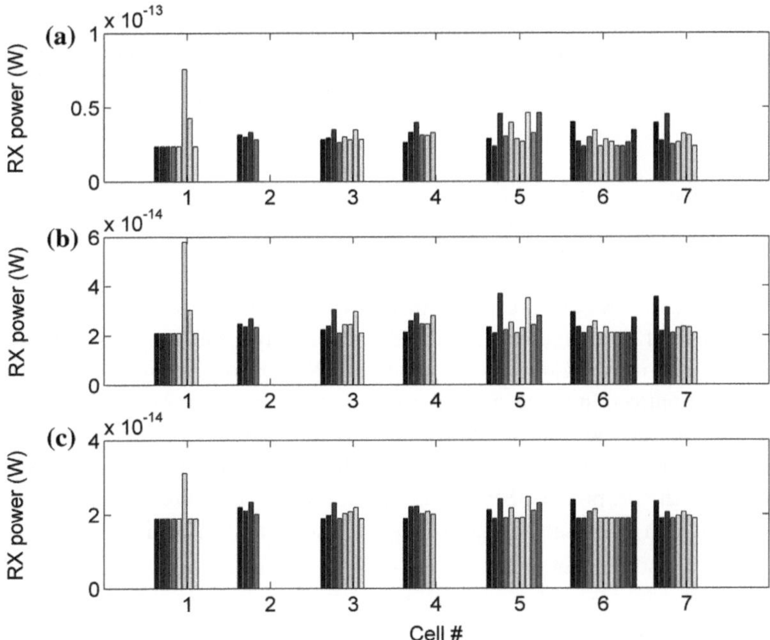

Fig. 5.8 A sample of sessions' received power at home BS ($M = 56$): **a** $\kappa = 4/7$ (*top*); **b** $\kappa = 9/14$ (*middle*); *c* $\kappa = 5/7$ (*bottom*)

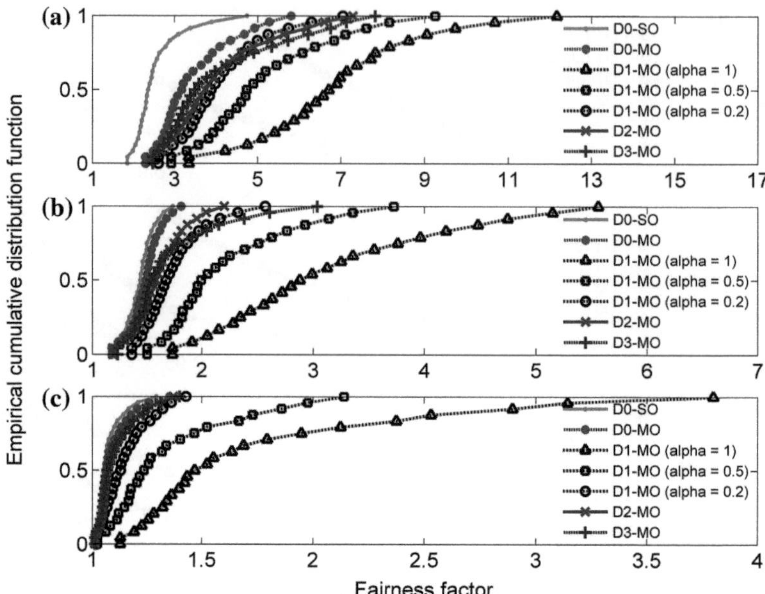

Fig. 5.9 Intra-cell fairness of USFR ($M = 56$): **a** $\kappa = 1/2$ (*top*); **b** $\kappa = 9/14$ (*middle*); **c** $\kappa = 6/7$ (*bottom*)

reduce the fairness factor as more subchannels become available, and thus improves intra-cell fairness.

5.6 Summary

In this chapter, the technique of USFR has been proposed in order to enhance the self-coexistence of CR networks operating in white-space (or unlicensed) spectrum. In view of the distributed nature of CR networks in a dynamic environment, we have resorted to the two-level game-theoretic approach that does not require direct inter-network coordination. In the non-cooperative game for multi-cell URA, the second-level TPC is solved optimally, while the first-level SCA is done heuristically in avoidance of perfect global knowledge and high computational complexity. Based on good heuristic, the proposed USFR technique has been shown to effectively improve self-coexistence in spectrum utilization, power consumption, and intra-cell fairness.

Chapter 6
Channel Assignment for Multi-hop Cognitive Radio Networks

Previous chapters study the coexistence problems in infrastructure-based CR networks, while this chapter has a focus on the coexistence issues in multi-hop CR networks. Specifically, for single radio interface (a half-duplex cognitive radio), multiple-hop CR networks, we propose a channel assignment strategy that uses the "segment" as the granularity for making channel assignment decisions. Nodes within intersecting flows (i.e., a component) can be grouped into one or more sets according to the set of channels available to each node—i.e., all nodes in the same segment have access to the same set of common channels. The channel assignment problem has an added dimension of complexity in the context of CR networks that does not exist in the context of conventional wireless networks. This complexity is the need to switch channels due to the temporal and spatial spectrum variability caused by incumbent users' spectrum utilization.

The proposed segment based channel assignment strategy deals with the spectrum variability problem by incorporating an adaptive segment maintenance mechanism that includes mechanisms for segment merging and segment splitting. During the lifetime of a flow, a given segment may need to split into smaller segments or merge as part of a larger segment due to the appearance/disappearance of primary user signals in the current operating band. The splitting/merging of segments leads to channel switching. Channel switching is one of the major factors leading to performance degradation in link based schemes. Practical limitations caused by channel switching include switching delay, scheduling overhead, and synchronization requirements. Although channel switching cannot be avoided completely in the context of CR networks, the proposed channel assignment scheme minimizes the need for it by using the segment as the granularity of channel assignment.

This chapter is organized as follows. We present a discussion of existing channel assignment strategies and the motivation for segment-based channel assignment in Sects. 6.1 and 6.2. In Sect. 6.3, we provide details on a distributed algorithm for supporting segment-based channel assignment. In Sect. 6.4, we show the simulation results, and we summarize this chapter in Sect. 6.5.

© Springer International Publishing Switzerland 2014
K. Bian et al., *Cognitive Radio Networks*,
DOI: 10.1007/978-3-319-07329-3_6

6.1 Channel Assignment Strategies

A taxonomy of channel assignment strategies is shown in Fig. 6.1. Not that, to avoid the interference to incumbent users in CR networks, we derive the segment-based channel assignment strategy from the component-based approach.

6.1.1 Link-Based Channel Assignment

All packets on a wireless link between two nodes are transmitted on the same channel until the channel assignment decision expires. Each link in a flow can choose any one of the free channels. Several approaches such as common control channel based channel negotiation [92, 113], D1EC [5], SSCH [6] and MMAC [92] fall under the category of link level assignment. Note that existing channel assignment strategies for CR networks take the link-based approach, such as [13, 15, 120, 121]. The major pitfall of a link-based approach is the significant channel switching delay incurred when a node serves two links on different channels. Therefore, they suffer from the practical limitations of the link-based strategy. Those limitations include switching delay, synchronization requirements, and scheduling overhead. Note that channel switching is required in the link-based approach when an intersection node serves two links in different channels.

6.1.2 Flow/Component-Based Channel Assignment

In flow-based channel assignment, all packets belonging to a flow are transmitted on the same channel. Different flows may operate on different channels. A graph is connected if there is a path connecting every pair of vertices. A graph that is not connected can be divided into connected components (disjoint connected subgraphs). A component in the context of channel assignment is similarly defined as a connected subgraph in the network flow graph, which is composed of nodes belonging to intersecting flows. In component-based channel assignment, all nodes within a component are assigned the same channel. If there are no intersecting flows, the component-based assignment is equivalent to the flow-based assignment.

This chapter is organized as follows. We present a discussion of motivation in Sect. 6.2. In Sect. 6.3, we provide details on a distributed algorithm for supporting segment-based channel assignment. In Sect. 6.4, we show the simulation results and summarize this chapter in Sect. 6.5.

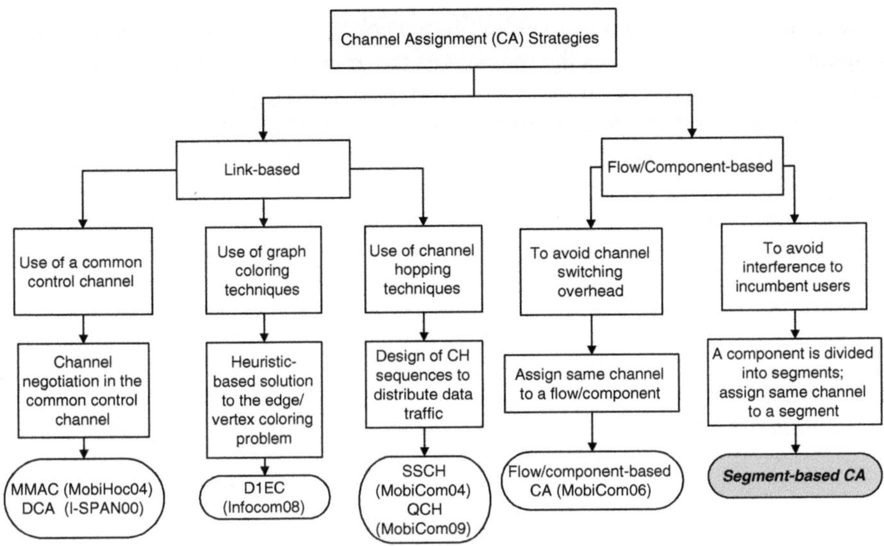

Fig. 6.1 A taxonomy of channel assignment strategies

6.2 Motivation for Segment-Based Channel Assignment

6.2.1 Segment-Based Channel Assignment

In conventional multi-channel wireless networks, all channels are fixed in terms of frequency and bandwidth [84]. In CR networks, the spectrum opportunities dynamically vary according to the primary users' transmissions. A flow that traverses through areas within the range of transmitting primary users experiences dynamically changing channel availability. Thus, it is not possible to assign the same channel to all the links within a flow or component due to the presence of primary signals. This means that flow- and component-based channel assignment algorithms used in conventional wireless networks cannot be applied in CR networks. For this reason, existing channel assignment schemes for CR networks use the link as the granularity of channel assignment. Unfortunately, the link-based approach incurs performance disadvantages due to frequent channel switching. Note that an intersection node that serves multiple links or flows in different channels needs to carry out channel switching. The aforementioned drawbacks include switching delay, synchronization requirement, and scheduling overhead. The latter two drawbacks can be mitigated by employing a control channel. This approach, however, is not desirable for single radio interface network if channel switching overhead is significant, as it requires frequent switching between the control channel and the data channels. Channel assignment approaches for multi-radio networks are discussed in [4].

To address the drawbacks of the link-based approach, we propose a new channel assignment strategy that uses the segment as the granularity of channel assignment. The proposed approach addresses the spectrum variability problem (caused by primary user transmissions) by considering spectrum sensing results in channel assignment decisions, and outperforms the link-based approach by requiring less frequent channel switches.

Before defining the terms segment and segment-based channel assignment, we need to define the term component. A component in a flow graph[1] is defined as a maximally connected sub-graph [110] such that there exists a path from any node in the sub-graph to all other nodes in the sub-graph. Next, we define the following terms:

Definition 8 A segment is defined as the maximal sub-graph of a component in which all nodes in the sub-graph have access to at least one common channel.

Definition 9 The two nodes at both ends of a link that connects two segments of the same component are defined as segment gateway nodes.

The segment-based channel assignment strategy assigns the same channel to all nodes within a segment while conforming to the spectrum utilization rules of the OSS paradigm. The assigned channel is called the operation channel of that segment. An illustration of the segment-based channel assignment strategy is shown in Fig. 6.2. Due to the spatial variability in spectrum opportunities, secondary users in different locations may detect different spectrum "white spaces". Suppose there are two primary user transmitters, PU1 and PU2, transmitting in channels 1 and 2, respectively. We assume that the primary network is a TV broadcast network. There is a multi-hop network composed of CR nodes, coexisting with those two primary users. There are two channels in total and two data flows: Flow 1 from node 0 to node 4 and Flow 2 from node 5 to node 6. Due to PU1 on channel 1 and PU2 on channel 2, the only available channel for nodes $\{0, 1, 5, 6\}$ is channel 2, while the only available channel for nodes $\{3, 4\}$ is channel 1. Node 2 is outside the interference range of the two primary transmitters, and hence has access to both channels. If Definition 8 is applied to the scenario given in Fig. 6.2, then one can see that nodes $\{0, 1, 5, 6\}$ form one segment and nodes $\{2, 3, 4\}$ form another segment. Note that node 2 can belong to either segment by Definition 8. In Fig. 6.2, nodes 1 and 2 are the gateway nodes.

In the above discussions, we have implicitly assumed that the channels of a primary network are equivalent to those of a secondary network. However, this may not be true. For example, the IEEE 802.22 standard [50] supports channel bonding such that up to 3 TV channels of a primary network can be bonded to one channel for use by the secondary network to support high-bandwidth applications. Since our assumption regarding the channels does not affect our discussions of the proposed segment-based channel assignment scheme, we will continue to make this assumption for the sake of simplicity.

---

[1] The set of active links carrying flow traffic in a network.

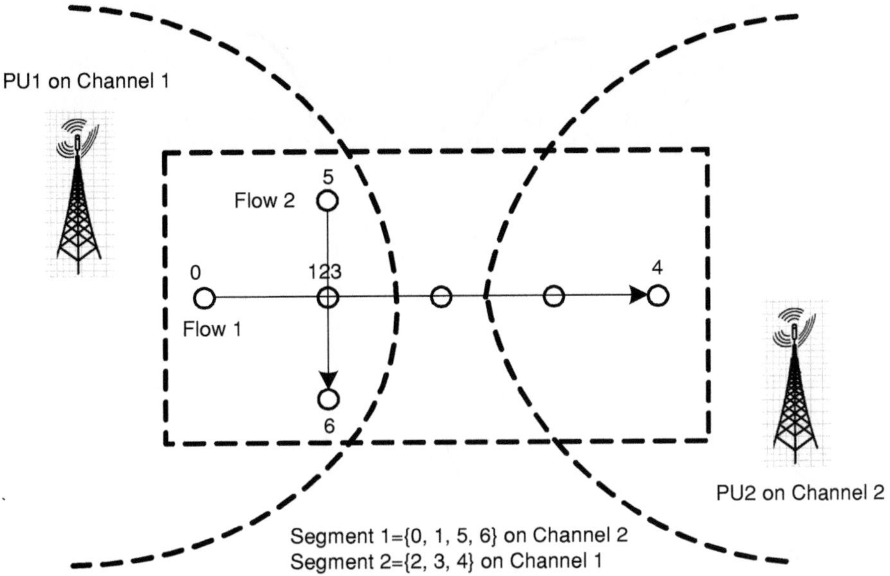

Fig. 6.2 An illustration of segment-based channel assignment

6.2.2 Performance Considerations

Link- and flow-based approaches require channel switching when a node serves two links or two flows on different channels. The segment-based approach requires channel switching when two segment gateway nodes on different channels need to communicate with each other. In the following discussions, we compare the segment-based approach with the link-based approach using flow capacity.

6.2.2.1 Flow Capacity Without Channel Switching

In the following analysis, we make the following assumptions: (1) a total of three channels; (2) switching delay is ignored; (3) a two-hop interference region for the secondary network; and (4) links within the same interference region in the same channel are assigned to different slots. We analyze two different cases.

Case 1: Contending flows. We compare the flow capacity of link- and segment-based approaches for the case of contending flows[2] Figure 6.3 shows the slot and channel assignments for two contending flows. The operation channels of the two primary users are also shown. We observe that it is possible to assign the slots in such

---

[2] Two flows are said to be contending, if there is at least one node in the set of active nodes of one flow that is within the interference region of the set of active nodes of the other flow.

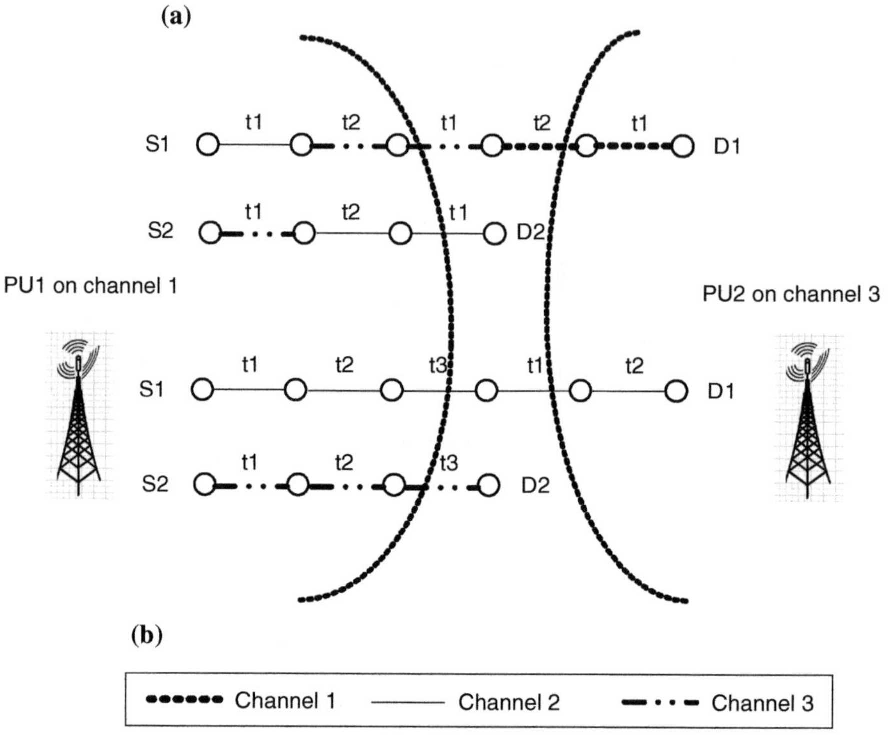

Fig. 6.3 Channel and slot assignments for contending flows

a way that links within the same contention region are assigned to different slots. In the following discussions, we assume that the link capacity is W.

Using the link-based channel assignment, the per-flow capacity is limited to $W/2$, irrespective of the number of channels and the slot schedule if we assume that each node is equipped with a single, half-duplex radio. In Fig. 6.3a, the two contending flows each achieve a per-flow capacity of by using channel and slot assignment schemes that utilize three channels and two slots, t_1 and t_2.

In the segment-based approach, a single flow may be divided into one or multiple segments. The per-flow capacity is limited by the segment that has the smallest capacity value within the flow. In Fig. 6.3b, there is only one segment in each flow. The maximum flow capacity of each contending flow is $W/3$, which is equal to the maximum per-flow capacity of a single flow using segment-based channel assignment.

From the above discussions, we can conclude that the per-flow capacity for both link- and segment-based channel assignment strategies is $O(W)$ for non-intersecting, contending flows.

Case 2: Intersecting flows. Figure 6.4 shows an example of intersecting flows. Node X is the intersection node. Note that when two flows intersect, the per-link

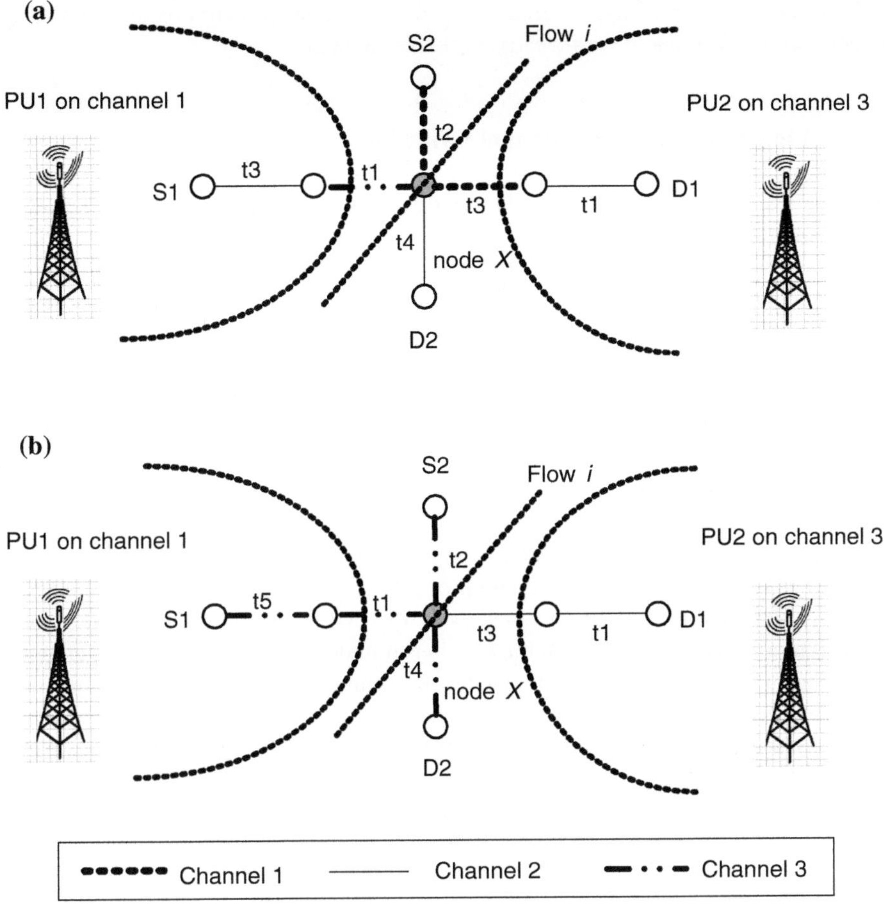

Fig. 6.4 Channel and slot assignments for intersecting flows **a** Link-based: flow capacity = W/4, **b** Segment-based: flow capacity = W/5

capacity is upper bounded by $W/4$ because the capacity of the links around the intersection node is limited to $W/4$ due to the fact that each node is equipped with a single half-duplex radio. In the link-based approach, the per-flow capacity is $W/4$ as shown in Fig. 6.4a. In the segment-based approach, we need to schedule five slots to avoid interference around the intersection node X. Hence, the per-flow capacity is $W/5$.

If we add one more flow (Flow i in Fig. 6.4) intersecting at node X, two more slots need to be scheduled. Suppose n flows intersect at node X, then the aggregate flow capacity of n flows under the link- and segment-based approaches are $W/2$ and $W/(2 + 1/n)$ respectively. From the above equations, we observe that the aggregate flow capacity of n intersecting flows under both channel assignment strategies is $O(W)$.

In the above discussions, we ignored the detrimental effects of channel switching on network performance; we investigate this issue next.

6.2.2.2 Flow Capacity with Channel Switching

When forwarding a packet, a secondary node needs to switch channels if two neighboring links are assigned to different channels. For a typical 802.11 network interface card, the switching delay is of the order of 80–100 μs and the transmission time for a 1 KB packet at 54 Mbps is 160 μs [105]. Thus, the switching delay in the aforementioned example is of the same order as the transmission time of a data packet. The switching delay contributes to the increase in the end-to-end delay of each packet's transmission as the switching delay is additive across all nodes that perform switching. According to [61], a wireless network's capacity C degrades as a function of $S/(S + T)$, where S is the switching delay and T is the transmission time. Suppose that there are x intermediate nodes in one flow that serve two neighboring links on different channels. Also, suppose that the flow is divided into y segments. The channel switching delay is t_s. The additive switching delay for link- and segment-based approaches is given by the following relations: $S_{link} = x \cdot t_s$ and $S_{segment} = (y-1) \cdot t_s$. Typically, the number of segments in a given flow (when using segment-based channel assignment) is much less than the number of nodes that need to perform channel switching (when using link-based channel assignment), thus $y < x$. Assuming other factors are the same, the capacity degradation caused by the switching delay in the link-based approach is greater than that in the segment-based approach. Thus, we can conclude that $C_{link} < C_{segment}$, where C_{link} represents the estimated capacity under link-based approach and $C_{segment}$ represents the estimated capacity under segment-based approach.

6.3 Segment-Based Channel Assignment

In this section, we present a distributed approach for realizing the segment-based channel assignment strategy. The proposed approach includes a channel assignment scheme and an adaptive segment maintenance scheme.

6.3.1 Initial Handshake

Before communicating with neighboring nodes, a secondary node needs to inform its neighbors about its current operation channel and the list of channels that is available to it. Each node in the network is required to broadcast an initial message (IM) on all of its available channels in a round-robin manner; this procedure is similar to the initial handshake procedure proposed in [120]. An alternative way for the initial handshake is based our proposed rendezvous protocols in Chap. 3. The IM of node i includes information regarding its available channel set A_i and its operation channel. Through the received IMs, a node acquires knowledge about the available channels and the operation channel of each neighboring node. Once the initial handshake is finished, the channel assignment process can start. Note that the proposed channel assignment scheme performs channel assignment, segment formation, and route discovery in an integrated manner. A node needs to broadcast an updated IM to its neighbors every time it changes its operation channel or revises its list of available channels.

6.3.2 Channel Assignment

The proposed channel assignment scheme can be integrated into any routing protocol for wireless multi-hop networks in which the source node of a flow initiates the route discovery process. Here, we take Dynamic Source Routing (DSR) [58] as an example.

6.3.2.1 Route Request Propagation

The proposed scheme's route request propagation phase is similar to the route request broadcast in DSR, and is described below.

1. A source node broadcasts a route discovery request (RREQ) message on all of the operation channels of its neighboring nodes. The operation channels of the neighbors are obtained from the IMs collected in the initial handshake phase.
2. When an intermediate node j receives the RREQ message, it piggybacks channel contention information on it and broadcasts the modified RREQ message to its neighbors. Specifically, $(cl_j(i_1), cl_j(i_2) \cdots)$ is piggybacked on the RREQ, where $cl_j(c)$ is the contention level of channel $c \in A_j$ perceived by the intermediate node j. The contention level of a channel is defined as the number of neighboring nodes that may potentially access this channel. The contention level value is calculated using the information in the IMs exchanged in the initial handshake procedure. As an example, let us calculate the value $cl_j(c)$. Suppose an intermediate node j has a set of neighbors represented by N_j, and each neighbor r has an available channel set A_r—both values can be obtained from the IMs. For $c \in A_j$ and $r \in N_j$, we define an indicator function as

$$I_j(c, A_r) = \begin{cases} 1, & \text{if } c \in A_r, \\ 0, & \text{otherwise.} \end{cases}$$

Then, the contention level of channel $c \in A_j$ perceived by node j can be calculated according to

$$cl_j(c) = \sum_{r \in N_j} I_j(c, A_r).$$

The channel contention level values are used in making channel selection decisions.

3. Using the received RREQs, the destination node constructs a set of disjoint paths and selects the shortest path from the set. Here, we assume the shortest-path (i.e., the path with the minimum hop number) selection policy. Then the destination node transmits a route reply (RREP) message back to the source via the reverse path, and prepares for channel assignment and segment formation.

6.3.2.2 Route Reply and Channel Assignment

In a network of secondary nodes, different portions of a route may use different channels due to channel availability influenced by spectrum variability. Suppose A_j denotes the list of available channels for node j. Then, each route has a corresponding available channel set, represented by $A_1, A_2, \ldots, A_k$, where k is the number of nodes on the route. In order for a route to be valid, relation (6.1) needs to be satisfied:

$$A_j \cap A_{j+1} \neq \emptyset, j = 1, \ldots, k - 1. \tag{6.1}$$

Relation (6.1) is satisfied when every pair of neighboring nodes on the route has at least one common channel available to both nodes. The RREP message transmitted by the destination node traverses through the k nodes towards the source node via the reverse-path.

1. Destination node k selects an available channel i^* with the smallest contention level value from set A_k, i.e.,

$$i^* = \arg\min_{i \in A_k} \sum_{j=1}^{k} cl_j(i). \tag{6.2}$$

Destination node k records its assigned channel as $ch_k = i^*$. The contention level values are obtained from the piggybacked information in the RREQs. Destination node k creates a new segment s_k and assigns itself to this segment. This new segment currently includes only one node, namely node k. The destination node piggybacks the 3-tuple (s_k, ch_k, A_k) in an RREP and sends it on the operation channel of the next node upstream[3] (i.e., node $(k - 1)$).

2. When an RREP (with the piggybacked tuple) is received on its operation channel, an intermediate node j checks to see whether it is already associated with an existing segment, s^*, and assigned to a channel, ch_{s^*}. i.e., it checks whether $s_j = s^*$ and $ch_j = ch_{s^*}$. If node j is already associated with a segment, it ignores the piggybacked information in the RREP. Then, node j notifies all the downstream nodes in its flow that are within the same segment about the channel assignment

---

[3] The upstream direction is the direction from the destination to the source. The downstream direction is the opposite of upstream.

and the segment association. After receiving the notification message, each notified node resets its assigned channel as ch_j and segment association as s_j. Next, node j piggybacks the tuple (s_j, ch_j, A_j) in the RREP and sends it to the next node upstream (i.e., node $(j-1)$) on the operation channel of node $(j-1)$.

3. Suppose that an intermediate node j has not committed to any existing segment and

$$ch_{j+1} = A_j, j = 1, \ldots, k-1. \tag{6.3}$$

That implies that node j can access the operation channel of the downstream neighbor node $(j+1)$. Thus, node j selects the same channel and associates itself with the same segment as node $(j+1)$, i.e., $ch_j = ch_{j+1}$ and $s_j = s_{j+1}$. Node j piggybacks the tuple (s_j, ch_j, A_j) in the RREP and sends it to the next upstream node (i.e., node $(j-1)$) on the operation channel of node $(j-1)$.

4. If node j is not committed to any existing segment and relation (6.3) does not hold, then node j creates a new segment, s_j, and selects a channel from A_j as its operation channel, ch_j. When selecting a channel, node j applies the same rule that the destination node applied in selecting its operation channel (see step (1) in Sect. 6.3.2.2). Node j piggybacks the tuple (s_j, ch_j, A_j) in the RREP and sends it to its upstream node.

5. When the RREP arrives at the source node, the source generates an ACK and sends it to downstream nodes of this route on their current operation channels to notify them that it has received the RREP. When a downstream node receives an ACK from the source, the operation channel and segment association selected by this node take effect. In Fig. 6.5, we have applied the aforementioned channel assignment scheme to a simple two-segment network as an example.

Note that our protocol uses the RREP messages to perform channel assignment as opposed to using RREQ messages (cf. [105]).

6.3.3 Segment Maintenance

The temporal changes in the primary users' utilization of the spectrum would cause changes in spectrum availability for the secondary users, thereby causing some of the secondary nodes to switch to a different channel. This, in turn, would induce changes in the segments within a secondary network. Two scenarios can occur: segment splitting and segment merging. In the following discussions, we discuss both scenarios.

6.3.3.1 Segment Splitting

The appearance of a primary user's signal in a spectrum band that is currently occupied by secondary nodes causes those secondary nodes to vacate that band and move to a fallow band by switching to another channel. Note that the appearance of a

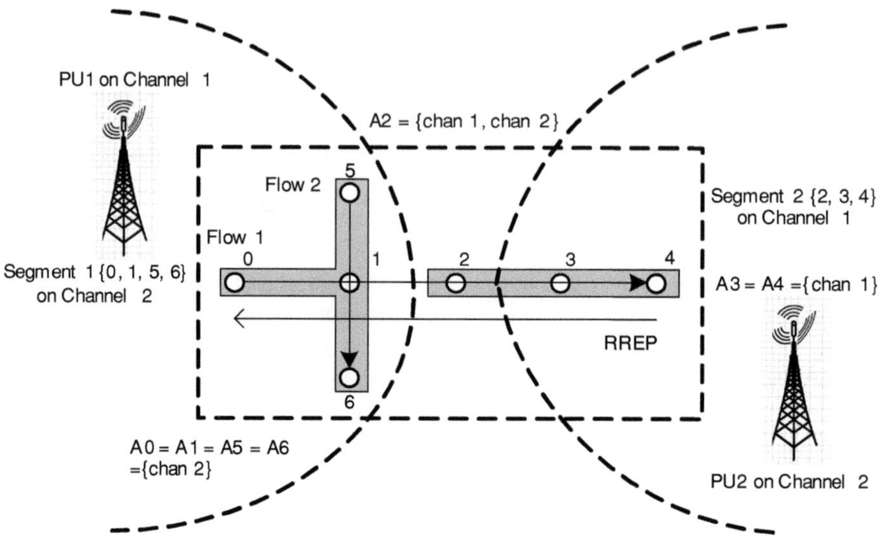

Fig. 6.5 Two segments are formed via the segment-based channel assignment approach

primary user does not necessarily require all of the nodes within a segment to switch channels; only a subset of the nodes may need to switch. Suppose that an intermediate node j detected the presence of a primary user signal and needs to switch to a different channel. Node j triggers a Channel Change message (CCHG) which is propagated towards the source (using the current operation channel). After receiving the CCHG, the source node generates a Route Repair message (RR) and sends it downstream via the current route. RR is forwarded towards the destination until it reaches node j^*. Node j^* is the node closest to the source node among the nodes that generated a CCHG. Node $j*$ initiates a route request propagation procedure to re-establish a new route to the destination. The procedure is the same as the one described in Sect. 6.3.2. This triggers another round of channel assignments for all nodes downstream of node j^*. This new channel assignment procedure will lead to the splitting of the current segment into two or more smaller segments.

6.3.3.2 Segment Merging

As noted in [105], the performance limitation of link- and flow-based channel assignment schemes is primarily due to the switching delay and overhead incurred when an intersection node serves two links/flows in different channels. Hence, minimizing the number of channel switchings can enhance performance significantly. In the context of the proposed segment-based channel assignment scheme, the number of channel switchings can be reduced when one takes advantage of a scenario in which an active primary user halts transmitting and releases a previously occupied band. By utilizing

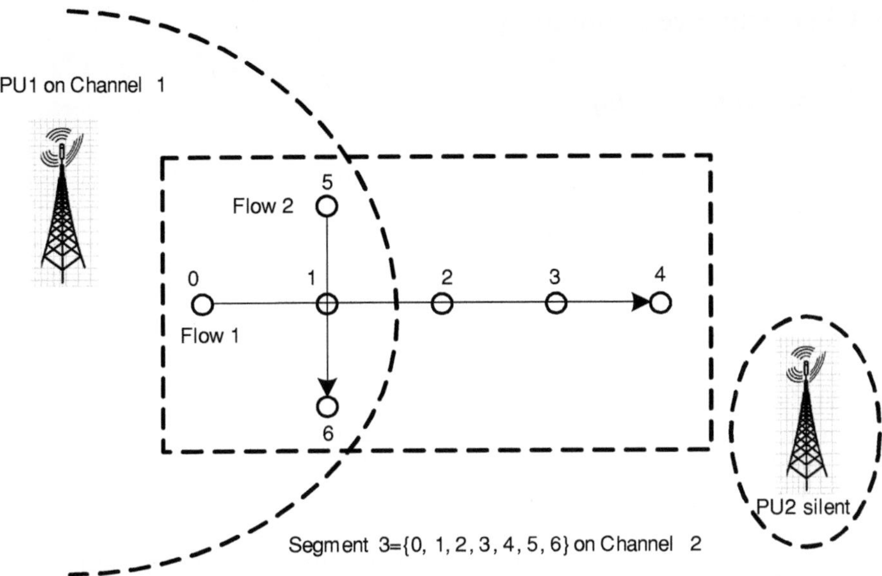

Fig. 6.6 Two segments merge into one if primary user 2 is silent

the recently released band, the secondary network may be able to reduce the number of segments, thus reducing the number of segment gateway nodes that need to carry out channel switching. An example is shown in Fig. 6.6. In this figure, Primary User 2 (PU2) has halted its transmission, thus releasing Channel 2, and enabling nodes 2, 3, and 4 to use Channel 2. Now, all nodes can be associated with a single segment that uses Channel 2. Comparing Figs. 6.5 and 6.6, we can see that the merging of the two segments into one will no longer require node 1 to carry out channel switching.

As noted above, the primary benefit of segment merging is the reduction of costs associated with channel switching (such as delay and overhead). However, segment merging is not always beneficial. Note that the segment merging process itself requires some nodes to perform a one-time channel switch. In fact, segment merging may even degrade performance when the period of the primary users' on/off cycle is short relative to the lifetime of a flow. For instance, if primary signals occupy and release the current band occupied by the segment's channel with high frequency, then multiple instances of segment merging and splitting may occur during the lifetime of a flow. In such a scenario, the overhead of segment merging/splitting would outweigh the benefits of segment merging, thus leading to poor performance. An interesting avenue for future research is investigating the tradeoffs of segment merging and its relationship to the performance of the secondary network.

6.4 Performance Evaluation

6.4.1 Simulation Setup

Throughout this section, we compare the proposed segment-based channel assignment scheme with the link-based channel assignment scheme. The simulations are carried out for a 1,000 m × 1,000 m square area with 50 secondary nodes placed randomly via a uniform distribution. All nodes are stationary. The transmission range of each secondary node is 250 m. We assume the existence of two primary signal transmitters—one transmitting in channel 1 and located at coordinates (0, 500), and the other one transmitting in channel 2 and located at coordinates (1,000, 500). The default channel capacity is 1Mbps, and the channel switching delay has a constant value of 100 μs. To simulate the proposed channel assignment scheme, we modified DSR to support channel assignments at the granularity of segments. In the simulated link-based approach, a link is assigned to a channel that has the smallest contention level value. For comparing the two channel assignment approaches, two evaluation metrics were used: average throughput and accumulative end-to-end delay. All results shown here are the results of averaging 10 simulation runs for each experiment.

6.4.2 Simulation Results

The simulation result shows superiority of our segment-based approach in performance over the link-base approach.

6.4.2.1 Invariant Primary User Transmission

We assume that the two primary users transmit during the entire duration of the simulation time interval. The total number of available channels is five. We compare the two channel assignment approaches using the average throughput of 10 UDP flows while varying the number of available channels. The results are shown in Fig. 6.7. We varied the total number of channels from 2 to 5.

6.4.2.2 Segment Merging in the Presence of Variant Primary User Transmission

To maintain an adequate level of performance, it is expected that the secondary network will choose to use licensed spectrum bands in which primary users' transmission pattern is not extremely dynamic (i.e., do not change very frequently). Thus, here we do not consider the case in which the primary users' transmission pattern varies frequently. We carried out a set of simulation experiments in which the pri-

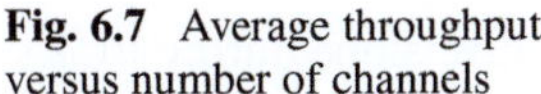

Fig. 6.7 Average throughput versus number of channels

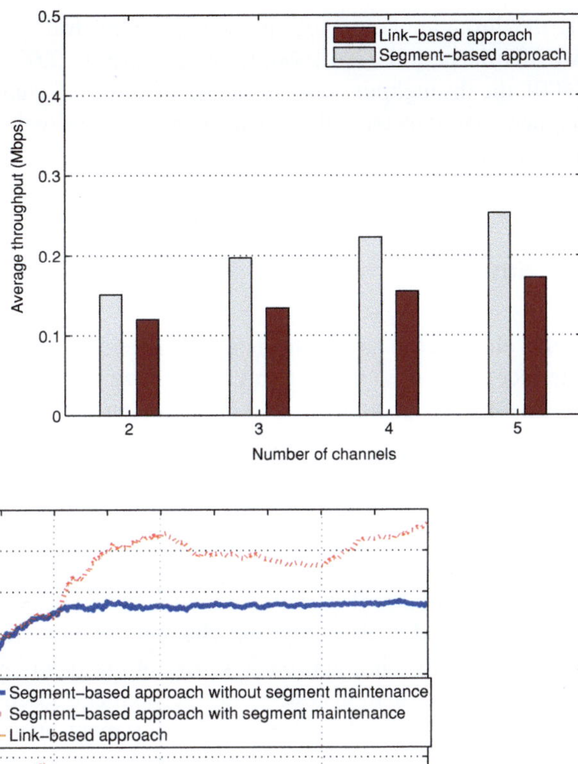

Fig. 6.8 Comparisons on the average throughputs of three protocols

mary user PU2 (operating in channel 2) has the following transmission schedule: it starts transmitting at the time point 0 sec; turns off at the 600-th sec; restarts transmitting at the 1000-th sec; and turns off again at the 1600-th sec. The total simulation time is 2000 sec. We compared the average throughput of three channel assignment schemes: segment-based approach without segment maintenance, segment-based approach with segment maintenance (i.e., segment merging and segment splitting), and link-based approach. The second approach performs segment merging when PU2 turns off to utilize the freed channel, and performs segment splitting when PU2 turns back on to avoid interfering with the primary signal transmission.

In Fig. 6.8, we observe that the segment merging process used by the second channel assignment scheme causes the average throughput to improve. We can clearly see the throughput curve of the second scheme follow a positive slope during PU2's two off time periods (i.e., [600, 1,000] and [1,600, 2,000]). During the interval [1,000, 1600], the throughput curve of the segment-based channel assignment scheme with segment maintenance follows a negative slope due to the delay and overhead involved in segment splitting.

6.5 Summary

In this chapter, we have considered the channel assignment problem in single radio interface, multi-hop CR networks. We have proposed a novel channel assignment scheme that assigns channels at the granularity of segments. In the OSS paradigm, the temporal and spatial spectrum variability caused by the primary users' spectrum utilization adds another dimension of complexity to the problem of channel assignment. Existing channel assignment approaches—such as link-based and component-based approaches—do not consider spectrum variability because they were originally designed for conventional wireless networks. Therefore, those approaches do not offer the best trade-off in terms of incumbent coexistence and network performance in the OSS model. The proposed channel assignment scheme is significantly simpler than the existing approaches, and offers practical benefits. Using simulation results, we have demonstrated that the segment-based channel assignment scheme outperforms the link-based channel assignment approach under realistic network conditions.

Chapter 7
Ecology-Inspired Coexistence of Heterogeneous Cognitive Radio Networks

Compared to the homogeneous coexistence problems studied in previous chapters, the coexistence of heterogeneous wireless systems in TV "white space" (TVWS) is a more challenging issue to address.

Owing the appealing characteristics of TVWS in wireless communications, industry and research stakeholders have initialized standardization efforts to enable the secondary networks' utilization of TVWS by leveraging cognitive radio (CR) technology. These efforts include IEEE 802.22 Wireless Regional Area Networks (WRAN) [50], IEEE 802.11af (WiFi over TVWS) [53], ECMA 392 (WPAN over TVWS) [24], etc. All of these standards rely on CR technology to overcome the challenging coexistence issues between primary and secondary networks as well as between secondary networks. In this chapter, we focus on the heterogeneous coexistence between secondary networks that employ different wireless technologies in TVWS, and we use the term "CR network" to denote a CR-enabled wireless network operating over TVWS.

The coexistence schemes for wireless networks can be broadly classified into two categories. A *non-collaborative* coexistence scheme is the only feasible approach when there are no means of coordination between the coexisting networks, such as the coexistence of WiFi and ZigBee networks [46, 123]. A *collaborative* coexistence scheme can be employed when coexisting networks can directly coordinate their operations, such as the self-coexistence schemes for 802.22 networks [12, 64].

Existing coexistence schemes fail to adequately address the heterogeneous coexistence problem in TVWS for a number of technical and policy reasons. Non-collaborative schemes cannot facilitate the coexistence among heterogeneous networks due to their incompatible MAC strategies. Collaborative strategies may require the exchange of potentially sensitive information (e.g., traffic load, bandwidth requirements) across different networks to negotiate the spectrum partitioning [109, 115], which could raise conflict-of-interest issues and customer privacy concerns for competing wireless networks or service providers. Moreover, it is difficult to find a third party that can serve as a global or centralized decision maker that supervise all heterogeneous networks and allocate spectrum them.

© Springer International Publishing Switzerland 2014

K. Bian et al., *Cognitive Radio Networks*,

DOI: 10.1007/978-3-319-07329-3_7

In this chapter, we propose a coexistence framework, called the *Symbiotic Heterogeneous coexistence Architecture (SHARE)*, that employs an *indirect* coordination method for enabling collaborative coexistence among heterogeneous CR networks. As its name implies, the proposed framework was inspired by the inter-species relations that exist in biological ecosystems. A *symbiotic* relation is a term used in biology to describe the coexistence of different species that form relations via indirect coordination. SHARE exploits a *mediator* system (e.g., the 802.19.1 system) that forwards *sanitized* data to establish the indirect coordination mechanism between coexisting networks. SHARE employs an *ecology-inspired spectrum sharing* algorithm inspired by an interspecific resource competition model that enables each CR network to autonomously determine the amount of spectrum that it should appropriate without direct negotiation with competing networks. Our analytical and simulation results show that SHARE guarantees weighted-fairness in partitioning spectrum and improves spectrum utilization.

In this chapter, we first outline a number of heterogeneous coexistence scenarios in section. We provide background knowledge of the mediator system and theoretical ecology in Sect. 7.2. In Sect. 7.3, we give an overview of SHARE. We present the SHARE algorithm and provide analytical results in Sect. 7.4. In Sect. 7.5, we evaluate the performance of SHARE using simulations. We conclude the chapter in Sect. 7.6.

7.1 Heterogeneous Coexistence Scenarios

Traditional wireless networks ranging from short-range personal or local area networks to long-range metropolitan area networks have their counterparts that support OSA in TVWSs. Hence, the HC problem existing in TVWSs among the new wireless standards that evolve from those traditional ones, including ECMA-392, 802.11af, 802.16h, and 802.22. Besides specific network systems, IEEE 802.19.1 [8] and COGEU project [77] are being developed to provide general solutions to the coexistence of 802 or non-802 networks in various CR-enabled use cases—e.g., campus, apartment complex, and home.

As the FCC issued a NPRM for enabling OSA in 3.5 GHz band [25, 26] lately, a total of 150 MHz spectrum that is being used by some critical federal and non-federal incumbents (e.g., radar and satellite services) has become a new focus of spectrum regulators and researchers. In the proposed framework, spectrum sharing can be achieved among three tiers of users, including tier-1 incumbent access users, tier-2 priority access users, and tier-3 general authorized access users. The services of tier-1 and tier-2 users should be protected from harmful interference in certain exclusion zones, while tier-3 users have to opportunistically access the band. All the users need to register in the spectrum access system (SAS), which incorporates a geolocation database and various interference mitigation techniques. The SAS as a centralized resource allocator is able to collect the location information of registered secondary users and utilize the information to achieve database-driven protection of primary users and coexistence of secondary users.

In general, the sharing of white space spectrum among coexisting networks can be achieved in several ways depending on whether or not decision-making coexistence infrastructures and inter-network coordination channels are available. The HC mechanisms can be classified into centralized, coordinated, and autonomous categories, as shown in Fig. 2.2. Centralized mechanisms require both coexistence infrastructures and coordination channels. A coexistence scenario for centralized mechanisms is illustrated in Fig. 2.2a. Each operator can deploy a single or multiple infrastructures to carry out centralized spectrum sharing via a star-topology or a cluster-based architecture. In each cluster, as in Fig. 2.2a, each coexistence infrastructure minimizes mutual interference among registered coexisting networks based on centrally collected coexistence information. Furthermore, multiple coexistence infrastructures can coordinate with each other for further collaboration. Coordinated mechanisms are applied when each coexisting network locally conducts resource allocation without the need for extra coexistence infrastructures but coordination channels are still available. A coexistence scenario that employs coordinated mechanisms is illustrated in Fig. 2.2b. The coexisting networks coordinate with each other through various information signaling and retrieving techniques. Based on the collected coexistence information, each network can make decisions to mitigate inter-network interference. Autonomous mechanisms are utilized by each coexisting network to achieve best-effort interference mitigation when neither coexistence infrastructures nor coordination channels are available. A coexistence scenario for autonomous mechanisms is illustrated in Fig. 2.2c. Each coexisting network performs resource allocation and manages inter-network interference only based on local observation.

7.2 Technical Background

As stated previously, SHARE employs a mediator system to establish an indirect coordination mechanism between CR networks. Note that the mediator is not a global decision maker. Using the sanitized information forwarded by the mediator, each CR network makes coexistence decisions autonomously using the algorithm proposed in this chapter.

7.2.1 The Mediator System

The recently formed IEEE 802.19.1 task group (TG) was chartered with the task of developing standardized methods, which are radio access technology-independent, for enabling coexistence among dissimilar or independently operated wireless networks [49]. This standard is currently being developed, and it has yet to prescribe solid solutions. The IEEE 802.19.1 system is a good candidate to serve as the mediator. The IEEE 802.19.1 system [49] defines a set of logical entities and a set of standardized interfaces for enabling coordination between heterogeneous CR net-

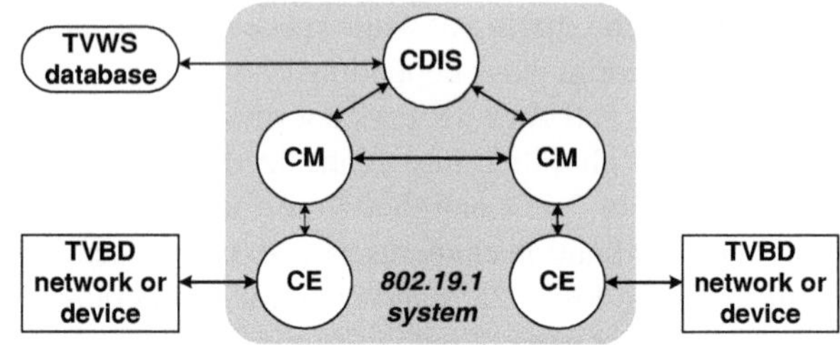

Fig. 7.1 IEEE 802.19.1 system architecture

works. In Fig. 7.1, we show the architecture of an 802.19.1 system which includes three entities in the grey box: (1) the coexistence manager (CM) acts as the local decision maker of the coexistence process; (2) the coexistence database and information server (CDIS) provides coexistence-related control information to the CMs, and (3) the coexistence enabler (CE) enables communications between the 802.19.1 system and the TV band device (TVBD) network. The TVWS database indicates the list of channels used by primary users and their locations, and it is connected to the 802.19.1 system via backhaul connections.

7.2.2 Interspecific Competition in Ecology

In ecology, interspecific competition is a distributed form of competition in which individuals of different species compete for the same resource in an ecosystem without direct interactions between them [101]. The impact of interspecific competition on populations have been formalized in a mathematical model called the Lotka-Volterra (L-V) competition model [67, 107]. In this model, the impact on population dynamics of species i can be calculated separately by a differential equation given below:

$$\frac{dN_i}{dt} = r_i N_i \left(1 - \frac{N_i + \sum_{j \neq i} \alpha_{ij} N_j}{K_i} \right). \tag{7.1}$$

In this equation, N_i is the population size of species i, K_i is the carrying capacity (which is the maximum population of species i if it is the only species present in the environment), r_i is the intrinsic rate of increase, and α_{ij} is the competition coefficient which represents the impact of species j's population growth on the population dynamics of species i. The interspecific competition model has been used for modeling the bandwidth allocation problem for TCP flows [41, 45].

7.3 Overview of the Protocol

In this section, we present the system model, underlying assumptions, and the architecture of SHARE.

7.3.1 System Model

We assume n heterogeneous networks are co-located, and they coexist in the same spectrum band that includes N channels with an identical bandwidth. Let $\mathcal{K}$ denote this set of networks, and all of these networks in $\mathcal{K}$ are registered with the mediator system. Every network is composed of multiple devices and a base station (BS) (or access point). Channels are labeled with indices $0, 1, \ldots, N - 1$.

Time-spectrum blocks. Time is divided into periods, each period contains a number of u super-frames, and each super-frame contains f frames (Such a structure based on frames can be found in IEEE 802.16 and 802.22). In this chapter, we use a time-spectrum block as the minimum unit for spectrum allocation, which can be defined by a channel index and a frame index. Specifically, we represent a time-spectrum block using a three-tuple (i, j, k)—i.e., the k-th frame in the j-th super-frame over channel i. Over channel i, there are a number of uf blocks that can be allocated during a period. We assume that a BS or network with multiple radios is able to scan and access multiple time-spectrum blocks on different channels simultaneously. Furthermore, we define the capacity, C, as the total number of spectrum-time blocks during a period, given N channels.

The bandwidth requirement. We define the *bandwidth requirement* of a network as the number of time-spectrum blocks that it needs to satisfy the QoS requirements of its traffic load. Let R_i denote the bandwidth requirement of network i.

The mediator-based indirect coordination. SHARE establishes a mediator-based *indirect* coordination mechanism between coexisting networks. There is no direct coordination between the coexisting networks, and they have to interact with each other by exchanging control information through a third-party mediator. Specifically, SHARE utilizes a CDIS (which is one of the components of an 802.19.1 system) as a mediator. Note that CDIS is not a global or centralized decision maker, but rather it is an information directory server with simple data processing capabilities.

Necessity of sanitized information. The mediator helps address conflict-of-interest issues and customer privacy concerns, which may arise when coexisting networks operated by competing service providers are required to exchange sensitive traffic information in order to carry out coexistence mechanisms. The mediator *sanitizes* the sensitive information received from the coexisting networks and then returns the sanitized information back to them. The coexisting networks execute their coordinated coexistence mechanisms using the sanitized data.

7.3.2 Ecology-Inspired Spectrum Allocation

As mentioned before, spectrum allocation among the coexisting networks through *direct* coordination may not be possible (due to a lack of infrastructure), may be too costly, or may be shunned by the competing network operators because they do not want to provide their sensitive information. Instead of direct coordination, the

SHARE framework adopts an indirect coordination mechanism, which is inspired by an interspecific competition model from theoretical ecology.

Design objective. In a spectrum sharing process, a network has to figure out how much spectrum it can appropriate given its bandwidth requirement. Suppose a time-spectrum block is the minimum unit amount of spectrum allocation. Let S_i denote the *number of time-spectrum blocks* allocated to network $i \in \mathcal{K}$. We refer to S_i as the *spectrum share* of network i.

Our objective is that the spectrum sharing process will eventually reach a state of equilibrium, where the number of allocated blocks to each network is proportional to its reported bandwidth requirement.

Inspiration from ecology. In ecology, the population dynamics of a species in the interspecific resource competition process can be captured by the L-V competition model. In the context of network coexistence, we build a *weighted* competition model to help a network to determine the *dynamics of its allocated spectrum*, given its bandwidth requirement.

Information exchange between the mediator and a network. The mediator exchanges two types of control information with every CR network:

1. *Upload of local report.* Network i reports the current value of S_i to the mediator;
2. *Download of sanitized data.* The mediator replies back to network i with the sanitized data, i.e., sum of numbers of time-spectrum blocks of all other coexisting networks, i.e., $\sum_{j \neq i, j \in \mathcal{K}} S_j$.

7.3.3 Problem Formulation

Suppose that $\mathcal{K}$ denotes a set of n co-located networks that have individual bandwidth requirements $R_1, R_2, \ldots, R_n$, and operate over the same WS. The first objective for coexisting networks is to split the WS into n pieces that are proportional to their individual bandwidth requirements, without sharing individual bandwidth requirements with each other.

Let $\mathbf{S}(\mathcal{K}) = [S_1, S_2, \ldots, S_n]$ denote the *spectrum share vector* for $\mathcal{K}$ over the WS.[1]

We define the *fairness index*, $F(\mathbf{S}(\mathcal{K}))$, for networks in $\mathcal{K}$ as follows:

$$F(\mathbf{S}(\mathcal{K})) = \frac{\left(\sum_{i \in \mathcal{K}} S_i\right)^2}{\sum_{i \in \mathcal{K}} R_i \cdot \sum_{i \in \mathcal{K}} R_i \left(\frac{S_i}{R_i}\right)^2}. \tag{7.2}$$

The maximum value of $F(\mathbf{S}(\mathcal{K}))$ is one (the best or weighted-fair case), where the allocated spectrum share value of a network is proportional to its bandwidth requirement.

---

[1] The vector is a row vector or a $1 \times n$ matrix.

Let $\mathcal{I}_i$ denote the *set of shared control information* known by network i, and it is easy to see that $R_i \in \mathcal{I}_i$. However, we assume that $R_j \notin \mathcal{I}_i$—i.e., co-located networks, i and j, do not know each other's bandwidth requirements.

We formulate a *weighted-fair spectrum sharing allocation* problem where heterogeneous networks dynamically determine their spectrum share values.

Problem 9 Given a set of n co-located networks, $\mathcal{K}$, operating over N channels, one has to solve the following problem to find the spectrum share vector for $\mathcal{K}$:

$$Maximize \ \ F(\mathbf{S}(\mathcal{K}))$$

$$subject \ to \ \ \frac{S_i}{S_j} = \frac{R_i}{R_j}, R_j \notin \mathcal{I}_i, \forall i, j \in \mathcal{K}.$$

The first constraint $\frac{S_i}{S_j} = \frac{R_i}{R_j}$ guarantees the weighted fairness.

7.4 An Ecology-Inspired Spectrum Share Allocation Algorithm

7.4.1 A Weighted-Fair Spectrum Competition Model

7.4.1.1 The Stable Equilibrium of the L-V Competition Model

The L-V competition model provides a method for defining a state of "stable equilibrium" and finding the sufficient conditions for achieving it. If one considers the interspecific competition process described by Eq. (7.1), when $K_i = K_j$ and $\alpha_{ij} = \alpha_{ji}$ for any two species i and j, then the sufficient condition for stable equilibrium is $\alpha_{ij} < 1$.

7.4.1.2 The Basic Spectrum Competition Model

In Table 7.1, we identify a number of analogies between a biological ecosystem and a network system. Based on Eq. (7.1) and the analogies, we can easily obtain the following *basic* spectrum competition model:

$$\frac{dS_i}{dt} = rS_i \left(1 - \frac{S_i + \alpha \sum_{j \neq i} S_j}{C} \right), \tag{7.3}$$

where S_i is the spectrum share for network i, and r is an intrinsic rate of increase. In Eq. (7.3), the carrying capacity is equal to the number of time-spectrum blocks in a period given N channels. A competition coefficient $\alpha < 1$ will guarantee a stable equilibrium—i.e., all the competing networks will have the same spectrum share value.

Table 7.1 A mapping between biological and CR network ecosystems

Biological ecosystem	CR network system
A species	A network
Population of a species	Spectrum share of a network
Population dynamics (growth or decline)	Dynamics of spectrum share

Next, we will show how to extend the basic competition model to a weighted-fair spectrum competition model that complies with the weighted-fairness requirement (i.e., $\frac{S_i}{S_j} = \frac{R_i}{R_j}$ for any two networks i and j) in a state of stable equilibrium.

7.4.1.3 The Weighted-Fair Spectrum Competition Model

The basic spectrum competition model guarantees a stable equilibrium where all the competing networks have the same spectrum share value. However, solutions to Problem 9 must satisfy the requirement of weighted fairness, which implies that the competing networks' spectrum share values are proportional to their bandwidth requirements. For example, if network i has a bandwidth requirement that is twice that of network j, then network i's allocated spectrum share should also be twice the allocated spectrum share of network j.

To support the weighted-fairness in spectrum share allocation, we construct a weighted-fair spectrum competition model by introducing the concept of "subspecies". A network with a higher bandwidth requirement would have a greater number of sub-species than a network with a lower bandwidth requirement. We use the bandwidth requirement R_i as the number of sub-species of network i.

Let $S_{i,k}$ denote the spectrum share allocated to the sub-species k of network i, where $k \in [1, R_i]$. In the *weighted* competition model, every sub-species k of network i calculates the change in its spectrum share according to the following equation:

$$\delta_{i,k} = \frac{dS_{i,k}}{dt}$$
$$= rS_{i,k}\left(1 - \frac{S_{i,k} + \alpha\sum_{\kappa \neq k} S_{i,\kappa} + \alpha\sum_{j \neq i} S_j}{C}\right). \tag{7.4}$$

Then, network i obtains its spectrum share value by combining the spectrum share values of all its sub-species, i.e., $S_i = \sum_k S_{i,k}$.

In SHARE, every network i periodically sends its spectrum share value S_i to the mediator, and then the mediator sends back the sanitized data $\beta_i = \sum_{j \neq i} S_j$ to network i. The spectrum share allocation process terminates when $\delta_{i,k} = 0$ for all i and k. Note that the sanitized data β_i is used (instead of actual bandwidth requirement information) to mitigate conflict of interest and privacy issues that may arise between competing networks. The use of sanitized data coincides with the second constraint of Problem 9. We show the pseudo code in Algorithm 7 and describe the procedure of SHARE as below.

1. A network i starts its spectrum share allocation process by creating a number of R_i sub-species.
2. At the beginning of every frame, every sub-species calculates the change rate of its spectrum share (i.e., $\frac{dS_{i,k}}{dt}$) using the sanitized data β_i obtained from the mediator.
3. If the change rate of the spectrum share is positive (or negative), a sub-species increases (or decreases) its spectrum share by randomly selecting a number of time-spectrum blocks to access (or releasing/freeing a number of occupied time-spectrum blocks).
4. At the end of every iteration, every network i calculates its new spectrum share value by $S_i = \sum_k S_{i,k}$, and sends S_i to the mediator. Meanwhile, the network updates the value of β_i from the mediator.
5. Last three steps are repeated until there is no sub-species with a non-zero change rate of spectrum share; that is $\frac{dS_{i,k}}{dt} = 0$ for every sub-species k of any network i.
6. The allocated spectrum share for network i is $\sum_k S_{i,k}$.

7.4.2 Characteristics of the Stable Equilibrium

Weighted-fairness. We first prove that the spectrum share allocation algorithm satisfies the requirement of weighted-fairness defined in Problem 9.

Lemma 5 *Given n coexisting CR networks in $\mathcal{K}$, when $\alpha < 1$, the spectrum share allocation process of Algorithm 7 is weighted-fair in partitioning the spectrum consisting of C time-spectrum blocks.*

Proof Suppose network $i \in \mathcal{K}$ has a number of R_i sub-species. The spectrum share allocation problem is equivalent to a problem where all sub-species compete for the resource using the L-V competition model. Since the sufficient condition for the equilibrium in the L-V competition model, $\alpha < 1$, is satisfied, the algorithm will terminate after a finite number of iterations, and all sub-species obtain the same spectrum share at the equilibrium point [41, 45], which is equal to $\frac{C}{\sum_{j \in \mathcal{K}} R_j}$. Hence, network i with R_i sub-species will obtain a spectrum share $R_i \frac{C}{\sum_{j \in \mathcal{K}} R_j}$, and thus $\frac{S_i}{S_{i'}} = \frac{R_i}{R_{i'}}, \forall i, i' \in \mathcal{K}$. $\qquad\square$

Stable equilibrium. We now show that the equilibrium point achieved by the weighted-fair competition model is stable.

Theorem 9 *Let $l = \sum_{i \in \mathcal{K}} R_i$ represent the total number of sub-species in the system. The differential Eq. (7.4) describe an l-dimensional system where the equilibrium when $S_i = R_i \frac{C}{l}$ is stable.*

Algorithm 7 The Spectrum Share Allocation Algorithm.

Input: competition coefficient α, capacity C, intrinsic rate of increase r, the sanitized data β_i.

Output: spectrum share, S_i, for network i.

1: Network i generates a number of R_i sub-species.
2: Update the value of β_i from the mediator.
3: **while** $\left(\exists k \in [1, R_i], s.t.\ \delta_{i,k} \neq 0\right)$ **do**
4:    **for** $k = 1$ to R_i **do**
5:      **if** $\delta_k \neq 0$ **then**
6:        $S_{i,k} = S_{i,k} + \delta_{i,k}$.
7:      **end if**
8:    **end for**
9:    Send $S_i = \sum_k S_{i,k}$ to the mediator and update the value of β_i.
10: **end while**
11: $S_i = \sum_k S_{i,k}$.

Proof Suppose networks in $\mathcal{K}$ generate a total number of l sub-species. For the sake of simplicity, we assign every sub-species an index from $\{1, \ldots, l\}$. Let $\mathbf{S}^* = [s_1^*, \ldots, s_l^*]$ be the spectrum share vector at the equilibrium point for all sub-species in the system, where s_i^* is the allocated spectrum share of sub-species i at the equilibrium point.

By Lemma 5, we have $s_i^* = \frac{C}{l}$, where $i \in [1, l]$. Equation (7.4) is equivalent to

$$\frac{ds_i^*}{dt} = rs_i^* \left(1 - \frac{s_i^* + \alpha \sum_{j \neq i, j \in [1,l]} s_j^*}{C} \right) = 0. \tag{7.5}$$

That is, $s_i^* + \alpha \sum_{j \neq i, j \in [1,l]} s_j^* = C$.

We will prove the equilibrium $\mathbf{S}^*$ is stable by linearizing the system equations at this equilibrium point. Let $\mathbf{S} = [s_1, \ldots, s_l]$ be a spectrum share vector for all sub-species at a non-equilibrium point. We denote the differential equation at this point as

$$G_i(\mathbf{S}) = rs_i \left(1 - \frac{s_i + \alpha \sum_{j \neq i, j \in [1,l]} s_j}{C} \right). \tag{7.6}$$

Let $\Delta s_i = s_i - s_i^*$. By linearizing equation (7.6) at the equilibrium point, we obtain

$$G_i(\mathbf{S}) = G_i(s_1^*, \ldots, s_l^*) + \sum_{i \in [1,l]} \left(\frac{\partial G_i(\mathbf{S})}{\partial s_i} \bigg|_{s_1^*, \ldots, s_l^*} \cdot \Delta s_i \right)$$

$$= -\left(\frac{r}{l} \right) \Delta s_i - \frac{r\alpha}{l} \sum_{j \neq i, j \in [1,l]} \Delta s_j. \tag{7.7}$$

We derive the l by l Jacobian matrix for the above Eq. (7.7) as follows

$$
A = \begin{vmatrix}
-\frac{r}{l} & -\frac{r\alpha}{l} & -\frac{r\alpha}{l} & \cdots & -\frac{r\alpha}{l} \\[6pt]
-\frac{r\alpha}{l} & -\frac{r}{l} & -\frac{r\alpha}{l} & \cdots & -\frac{r\alpha}{l} \\[6pt]
\vdots & \ddots & \ddots & \cdots & \vdots \\[6pt]
-\frac{r\alpha}{l} & -\frac{r\alpha}{l} & \cdots & -\frac{r\alpha}{l} & -\frac{r}{l}
\end{vmatrix},
$$

which is a symmetric matrix. This matrix has two eigenvalues $\lambda = -\frac{r}{l} - \frac{(l-1)r\alpha}{l}$ and $\frac{r(\alpha-1)}{l}$. Since $0 < \alpha < 1$, the two eigenvalues are negative. Based on the stability theory, the system is stable if all eigenvalues are negative. Hence, the differential equations shown by (7.4) describe an l-dimensional system and the equilibrium $\mathbf{S}^* = \{s_i^* | s_i^* = \frac{C}{l}, \forall i \in [1, l]\}$ is stable. $\square$

Convergence time. Next, we analyze the time required for the proposed algorithm to converge to the stable equilibrium.

Theorem 10 *Consider N networks that compete for the same spectrum band, then the time-to-convergence to the SHARE's equilibrium is $T_c = O(ln(C/l))$.*

Proof Similar to the proof of Theorem 9, there are a total number of l sub-species. Let $A = \sum_{j \neq i, j \in [1,l]} s_j = (l-1)s_0$, and Eq. (7.5) can be rewritten as

$$
\frac{ds_i}{dt} = rs_i \left(1 - \frac{s_i + \alpha A}{C}\right) = 0. \tag{7.8}
$$

By integrating (7.8), we can obtain

$$
s_i(t) = \frac{s_0 e^{rt\left(1 - \frac{\alpha A}{C}\right)}(C - \alpha A)}{s_0 \left(e^{rt\left(1 - \frac{\alpha A}{C}\right)} - 1\right) + (C - \alpha A)}. \tag{7.9}
$$

To calculate the time-to-convergence, we consider the time which is required to increase the spectrum share for network i from s_0 to $s_i(t) = s^* = C/l$. By solving (7.9), the time T_c becomes:

$$
T_c = \frac{C}{r(C - \alpha A)} ln \left(\frac{s_i(t)(C - \alpha A - s_0)}{s_0(C - \alpha A - s_i(t))}\right).
$$

The time of convergence of SHARE is $O(ln(C/l))$, and it is exponentially fast. $\square$

7.5 Performance Evaluation

In this section, we evaluate the performance of SHARE in two steps. We first look into the stable equilibrium achieved by the weighted-fair spectrum share allocation scheme. Then, we compare the foraging-based channel selection scheme and the random channel selection strategy in terms of system fitness.

7.5.1 The Stable Equilibrium

In the first set of simulations, we simulate two CR networks that coexist in a block of spectrum that is divided into 20 channels. We fix the bandwidth requirements of the two networks as $R_1 = 2$ and $R_2 = 3$, which implies that network 1 has two sub-species and network 2 has three in the spectrum share allocation process. In the L-V competition model, the competition coefficient $\alpha < 1$ and the intrinsic rate of increase $r < 2$ [76]. The discussions on how to choose appropriate parameter values to achieve fast convergence to an equilibrium can be found in [41, 76]. In this set of simulations we used $\alpha = 0.9$ and $r = 1.95$. Next, we show that the coexisting networks under SHARE achieve an equilibrium, where the spectrum share of each network is proportional to its bandwidth requirement.

Convergence to an equilibrium. From Fig. 7.2, we observe the dynamics of the spectrum share value of each network and each sub-species within a network. "Sub-species (i, j)" in the figure legend represents sub-species j within network i. The system converges to an equilibrium state in finite time where all sub-species of every network are allocated the same spectrum share value. The aggregate spectrum share value allocated to a network is proportional to its bandwidth requirement.

Stability of the equilibrium. To test the stability of the equilibrium point, we introduce two types of disturbance in bandwidth requirement by: (1) silencing the sub-species $(2, 3)$ for a short time period (from the 120 to 140th iteration), and (2) deleting the sub-species $(2, 3)$ at the 360th iteration. Figure 7.3 shows the dynamics of spectrum share values when the disturbance is introduced. As can be seen in the figure, the disturbance causes the system to deviate away from equilibrium, but the coexisting networks quickly converge to a new equilibrium point where the allocated spectrum share values are proportional to the new values of bandwidth requirements.

7.5.2 Weighted Fairness

We vary the number of coexisting CR networks, and in each simulation run, the bandwidth requirement, R_i, of each network i is randomly chosen from the range $[1, 5]$. We compare SHARE with a "non-collaborative" allocation scheme where every coexisting network determines its spectrum share value without coordinating

Fig. 7.2 Convergence to the equilibrium

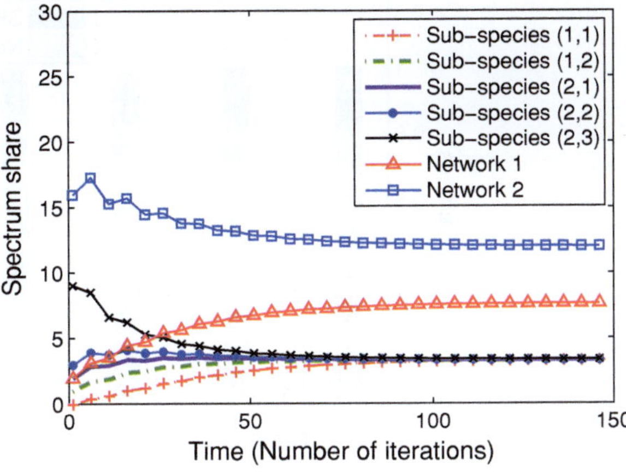

Fig. 7.3 Stability of the equilibrium

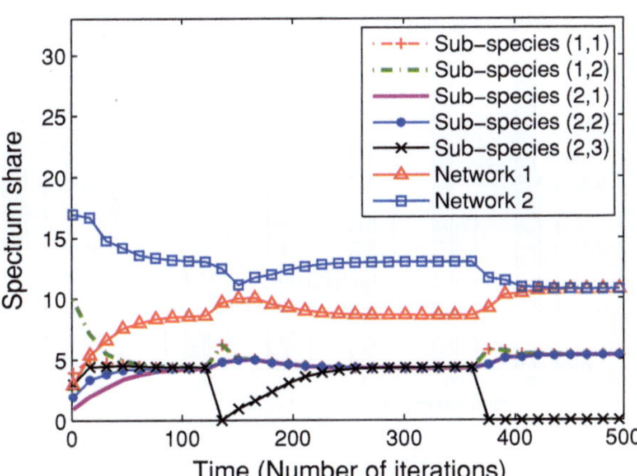

with others. This is equivalent to splitting the available spectrum "randomly" to n pieces and allocates them to n coexisting networks. We measure the fairness values using the fairness index defined in (7.2). Figure 7.4 clearly shows that SHARE allocates spectrum in a weighted-fair manner, whereas the non-collaborative allocation scheme does not.

7.5.3 System Satisfaction

We define the *satisfaction* of network i as the ratio between its allocated spectrum share to its bandwidth requirement, $f_i = \frac{S_i}{R_i}$. Then, the system satisfaction is the network satisfaction value of the network that has the minimum satisfaction, such as

$$\Phi = \min\{f_0, f_1, \ldots, f_{P-1}\}.$$

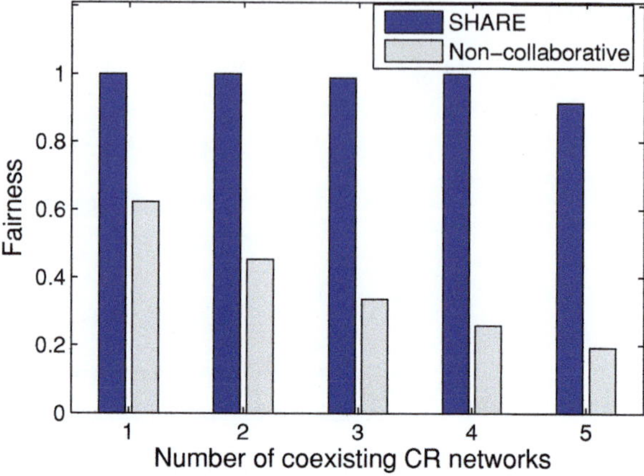

Fig. 7.4 Measured fairness values

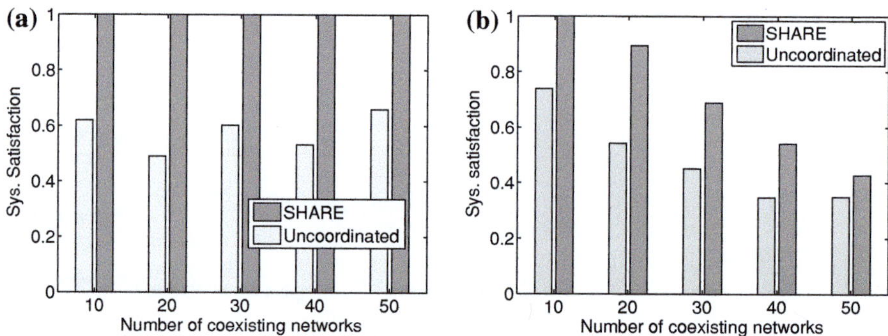

Fig. 7.5 System satisfaction: **a** coexisting networks have identical bandwidth requirement given sufficient spectrum; **b** coexisting networks have random values of bandwidth requirement given insufficient spectrum

Then, we vary the number of networks when their bandwidth requirement values are identical or uniformly distributed in a range $[1, N]$. From simulation results in Fig. 7.5, we observe that SHARE's system satisfaction value is close to one when the bandwidth requirement values are identical; when random bandwidth requirements are employed, SHARE's performance is much better that the random spectrum sharing strategy.

A centralized is always the best strategy when global information is available. However, in this chapter, due to incompatible interface, conflict of interests, privacy issues, we cannot deploy a centralized approach. Instead, we compare a mediator-based approach with the totally uncoordinated methods (the random strategy).

7.6 Summary

Inspired by symbiotic coexistence in ecology, in this chapter we presented a framework called Symbiotic Heterogeneous coexistence ARchitecturE (SHARE), which enables collaborative coexistence among heterogeneous CR networks over TVWS. SHARE enables two heterogeneous CR networks to coexist in TVWS through a mediator-based *indirect* coordination mechanism between them, which avoids the drawbacks of direct coordination mechanisms. The SHARE framework adopts two algorithms that are executed by each coexisting network to autonomously complete the following two spectrum sharing tasks: (1) dynamically determine its spectrum share that is proportional to its bandwidth requirement, and (2) select channels in such a way to achieve a very high value of system fitness. Analytical and simulation results show that SHARE enables the networks' spectrum allocation to converge to a stable equilibrium, and that in this allocation, weighted-fairness is ensured and the system fitness is maximized.

Bibliography

1. I.F. Akyildiz, W.Y. Lee, M.C. Vuran, S. Mohanty, Next generation/dynamic spectrum access/cognitive radio wireless networks: a survey. Comput. Netw. J. (Elsevier) **50**(13), 2127–2159 (2006)
2. M. Al-Shalash, F. Khafizov, Z. Chao, in *Proceedings of IEEE PIMRC'10*, Interference Constrained Soft Frequency Reuse for Uplink ICIC in LTE Networks, Sept 2010
3. M. Al-Shalash, F. Khafizov, Z. Chao, in *Proceedings of GLOBECOM'10*, Uplink Inter-Cell Interference Coordination through Soft Frequency Reuse, Dec 2010
4. M. Alicherry, R. Bhatia, L.E. Li, in *Proceedings of ACM MobiCom*, Joint Channel Assignment and Routing for Throughput Optimization in Multi-Radio Wireless Mesh Networks, pp. 58–72, Sept 2005
5. E. Aryafar, O. Gurewitz, E.W. Knightly, in *Proceedings of IEEE INFOCOM*, Distance-1 Constrained Channel Assignment in Single Radio Wireless Mesh Networks, pp. 762–770, Apr 2008
6. P. Bahl, R. Chandra, J. Dunagan, in *Proceedings of ACM MobiCom*, SSCH: Slotted Seeded Channel Hopping for Capacity Improvement in IEEE 802.11 Ad Hoc Wireless Networks, pp. 216–230, Sept 2004
7. P. Bahl, R. Chandra, T. Moscibroda, R. Murty, M. Welsh, in *Proceedings of Sigcomm*, White Space Networking with Wi-Fi like Connectivity, pp. 27–38, Aug 2009
8. T. Baykas, M. Cummings, H. Kang, M. Kasslin, J. Kwak, R. Paine, A. Reznik, R. Saeed, S.J. Shellhammer, Developing a standard for TV white space coexistence: technical challenges and solution approaches. IEEE Wirel. Comm. **19**(1), 10–22 (2012)
9. S. Buzzi, G. Colavolpe, D. Saturnino, A. Zappone, Potential games for energy-efficient power control and subcarrier allocation in uplink multicell OFDMA systems. IEEE J. Sel. Topi. Sign. Proc. **6**(2), 89–103 (2012)
10. K. Bian, in *Proceedings of IEEE INFOCOM, Mini-conference*, Asynchronous Channel Hopping for Establishing Rendezvous in Cognitive Radio Networks, Apr 2011
11. K. Bian, J. Park, R. Chen, in *Proceedings of ACM MobiCom*, A Quorum-Based Framework for Establishing Control Channels in Dynamic Spectrum Access Networks, pp. 25–36, Sept 2009
12. S. Brahma, M. Chatterjee, in *Proceedings of IEEE Globecom*, Mitigating Self-Interference Among IEEE 802.22 Networks: A Game Theoretic Perspective, pp. 1–6, Nov 2009
13. V. Brik, E. Rozner, S. Banarjee, P. Bahl, in *Proceedings of IEEE DySPAN*, DSAP: A Protocol for Coordinated Spectrum Access, pp. 611–614, Nov 2005

© Springer International Publishing Switzerland 2014

K. Bian et al., *Cognitive Radio Networks*,

DOI: 10.1007/978-3-319-07329-3

14. D. Cabric, S.M. Mishra, R.W. Brodersen, in *Proceedings of the 38th Asilomar Conference on Signals, Systems, and Computers*, Implementation Issues in Spectrum Sensing for Cognitive Radios, Nov 2004
15. L. Cao, H. Zheng, in *Proceedings of IEEE Second Annual IEEE Communications Society Conference on Sensor and Ad Hoc Communications and Networks, (Secon)*, DSAP: Distributed Spectrum Allocation via Local Bargaining, pp. 475–486, Sept 2005
16. R. Chen, J. Park, K. Bian, in *Proceedings of IEEE Infocom Mini-conference*, Robust Distributed Spectrum Sensing in Cognitive Radio Networks, Apr 2008
17. R. Chen, J. Park, J.H. Reed, Defense against primary user emulation attacks in cognitive radio networks. IEEE J. Sel. Areas Commun. Spec. Issue Cognitive Radio Theory Appl. **26**(1), 25–37 (2008)
18. R. Chen, J. Park, Y.T. Hou, J.H. Reed, Toward secure distributed spectrum sensing in cognitive radio networks. IEEE Commun. Mag. Spec. Issue Cognitive Radio Commun. **46**(4), 50–55 (2008)
19. T. Chen, H. Zhang, G.M. Maggio, I. Chlamtac, in *Proceedings of IEEE DySpan*, CogMesh: A Cluster-based Cognitive Radio Network, pp. 168–178, Apr 2007
20. C.J. Colbourn, E.J.H. Dinitz, *The CRC Handbook of Combinatorial Designs* (CRC Press, Boca Raton, 1996)
21. C. Cordeiro, K. Challapali, in *Proceedings of IEEE DySpan*, C-MAC: A Cognitive MAC Protocol for Multi-Channel Wireless Networks, pp. 147–157, Apr 2007
22. C.M. Cordeiro, K. Challapali, D. Birru, N.S. Shankar, in *IEEE DySPAN*, IEEE 802.22: The First Worldwide Wireless Standard Based on Cognitive Radios, Nov 2005
23. L.A. DaSilva, I. Guerreiro, in *Proceedings of IEEE DySpan*, Sequence-Based Rendezvous for Dynamic Spectrum Access, pp. 1–7, Oct 2008
24. Ecma International, *ECMA-392: MAC and PHY for Operation in TV White Space* 1st edn., Dec 2009
25. FCC, *Enabling Innovative Small Cell Use in 3.5 GHz Band NPRM & Order*. GN Docket No. 12–354, FCC 12–148, Dec 2012
26. FCC, *Wireless Telecommunications Bureau and Office of Engineering and Technology Call for Technical Papers on the Proposed Spectrum Access System for the 3.5 GHz Band*. GN Docket No. 12–354, DA 13–2213, Nov 2013
27. Federal Communications Commission, *Facilitating Opportunities for Flexible, Efficient, and Reliable Spectrum Use Employing Spectrum Agile Radio Technologies*. ET Docket No. 03–108, Dec 2003
28. Federal Communications Commission, *Unlicensed Operation in the TV Broadcast Bands and Additional Spectrum for Unlicensed Devices below 900 MHz in the 3GHz Band*. ET Docket No. 04–186, May 2004
29. R. Ferrus, O. Sallent, R. Agusti, Interworking in heterogeneous wireless networks: comprehensive framework and future trends. IEEE Wirel. Comm. **17**(2), 22–31 (2010)
30. F. Fitzek, D. Angelini, G. Mazzini, M. Zorzi, Design and performance of an enhanced IEEE 802.11 MAC protocol for multihop coverage extension. IEEE Wirel. Commun. **10**(6), 30–39 (2003)
31. B. Gao, J. Park, Y. Yang, S. Roy, A Taxonomy of coexistence mechanisms for heterogeneous cognitive radio networks operating in TV white spaces. IEEE Wirel. Commun. **19**(4), 41–48 (2012)
32. G. Ganesan, Y. Li, in *Proceedings of IEEE DySPAN*, Cooperative Spectrum Sensing in Cognitive Radio Networks, pp. 137–143, Nov 2005
33. V. Gardellin, S. K. Das, L. Lenzini, in *Proceedings of IEEE INFOCOM'10*, A Fully Distributed Game Theoretic Approach to Guarantee Self-Coexistence among WRANs, May 2010
34. V. Gardellin, L. Lenzini, V. Fontana, in *Proceedings of ACM CoRoNET'11*, Aware Channel Assignment Algorithm for Cognitive Networks, Sept 2011
35. S. Geirhofer, L. Tong, B.M. Sadler, in *Proceedings of IEEE GLOBECOM'08*, A Cognitive Framework for Improving Coexistence among Heterogeneous Wireless Networks, Nov 2008

36. M. Ghaderi, A. Sridharan, H. Zang, D. Towsley, R. Cruz, in *Proceedings of ACM MobiCom*, TCP-Aware Resource Allocation in CDMA Networks, pp. 215–226, Sept 2006
37. C. Ghosh, S. Roy, D. Cavalcanti, Coexistence challenges for heterogeneous cognitive wireless networks in TV white spaces. IEEE Wirel. Comm. **18**(4), 32–40 (2011)
38. D. Grandblaise, W. Hu, *Inter Base Stations Adaptive on Demand Channel Contention for IEEE 802.22 WRAN Self Coexistence*. IEEE docs: IEEE 802.22-07/0024r0, Jan 2007
39. D. Gurney, G. Buchwald, L. Ecklund, S. Kuffner, J. Grosspietsch, in *Proceedings of IEEE DySPAN'08*, Geo-Location Database Techniques for Incumbent Protection in the TV White Space, Oct 2008
40. Z. Han, Z. Ji, K.J.R. Liu, Non-cooperative resource competition game by virtual referee in multi-cell OFDMA networks. IEEE J. Sel. Area Comm. **25**(6), 1079–1090 (2007)
41. G. Hasegawa, M. Murata, *TCP Symbiosis: Congestion Control Mechanisms of TCP Based on Lotka-Volterra Competition Model*. ACM Inter-Perf, Oct 2006
42. S. Haykin, Cognitive radio: brain-empowered wireless communications. IEEE J. Sel. Areas Commun. Spec. Issue Cognitive Radio Theory Appl. **23**(2), 201–220 (2005)
43. Z. Hou, Y. Cai, D. Wu, Resource allocation based on integer programming and game theory in uplink multi-cell cooperative OFDMA systems. EURASIP J. Wirel. Comm. Netw. **2011**, 169 (2011)
44. S. Huang, X. Liu, Z. Ding, in *Proceedings of IEEE INFOCOM*, Opportunistic Spectrum Access in Cognitive Radio Networks, pp. 1427–1435, Apr 2008
45. X. Huang, F. Ren, G. Yang, Y. Wu, in *IEEE ICDCS*, End-to-end Congestion Control for High Speed Networks Based on Population Ecology Models, pp. 353–360, June 2008
46. J. Huang, G. Xing, G. Zhou, R. Zhou, in *Proceedings of IEEE ICNP*, Beyond Co-existence: Exploiting WiFi White Space for ZigBee Performance Assurance, Oct 2010
47. Huawei. Soft Frequency Reuse Scheme for UTRAN LTE. 3GPP R1–050507, Technical Report from Huawei, May 2005
48. IEEE 802.16 WG. IEEE Standard for Local and Metropolitan Area Networks, Part 16: Air Interface for Broadband Wireless Access Systems, Amendment 2: Improved Coexistence Mechanisms for License-Exempt Operation. IEEE Std., July 2010
49. IEEE 802.19 Task Group 1. Wireless Coexistence in the TV White Space, http://www.ieee802.org/19/pub/TG1.html
50. IEEE 802.22 Working Group, http://www.ieee802.org/22/
51. IEEE 802.22 Working Group, Reviews of Channel Model. IEEE docs: 22–05-0070-00-0000, Aug 2005
52. IEEE 802.22 Working Group, ETRI FT Philips Samsung Proposal. IEEE docs: 22–06-0005-01-0000, Jan 2006
53. IEEE P802.11 task group af. Wireless LAN in the TV White Space, http://www.ieee802.org/11/Reports/tgaf_update.htm/
54. J. Jia, Q. Zhang, X.S. Shen, HC-MAC: a hardware-constrained cognitive MAC for efficient spectrum management. IEEE J. Sel. Areas Commun. Spec. Issue Cognitive Radio Theory Appl. **26**(1), 106–117 (2008)
55. H. Jiang, S.S. Rappaport, CBWL: a new channel assignment and sharing method for cellular communications systems. IEEE Trans. Veh. Technol. **43**(2), 313–321 (1994)
56. J.R. Jiang, Y.C. Tseng, T. Lai, Quorum-based asynchronous power-saving protocols for IEEE 802.11 ad hoc networks. ACM J. Mob. Netw. Appl. **10**(1–2), 169–181 (2005)
57. X. Jing, D. Raychaudhuri, in *Proceedings of IEEE DySPAN'05*, Spectrum Coexistence of IEEE 802.11b and 802.16a Networks Using the CSCC Etiquette Protocol, Nov 2005
58. D.B. Johnson, D.A. Maltz, Y.-C. Hu, The Dynamic Source Routing Protocol for Mobile Ad Hoc Networks (DSR) (Internet-Draft). Mobile Ad-hoc Network (MANET) Working Group, IETF, July 2004
59. H. Kim, K.G. Shin, Efficient discovery of spectrum opportunities with MAC-layer sensing in cognitive radio networks. IEEE Trans. Mob. Comput. **7**(5), 533–545 (2008)
60. H. Kim, K.G. Shin, in *Proceedings of IEEE INFOCOM'10*, Asymmetry-Aware Real-Time Distributed Joint Resource Allocation in IEEE 802.22 WRANs, Mar 2010

61. P. Kyasanur, N. Vaidya, in *Proceedings of ACM MobiCom*, Capacity of Multi-channel Wireless Networks: Impact of Number of Channels and Interfaces, pp. 43–57, Sept 2005

62. L. Lazos, S. Liu, M. Krunz, in *Proceedings of ACM WiSec*, Mitigating Control-Channel Jamming Attacks in Multi-channel Ad Hoc Networks, pp. 169–180, Mar 2009

63. J.R. Leigh, *Essentials of Nonlinear Control Theory* (Peter Peregrinus Ltd., London, 1983)

64. C. Ko, H. Wei, Game theoretical resource allocation for inter-BS coexistence in IEEE 802.22. IEEE Trans. Veh. Technol. **59**(4), 1729–1744 (2010)

65. H. Kwon, B.G. Lee, in *Proceedings of IEEE ICC'06*, Distributed Resource Allocation through Noncooperative Game Approach in Multi-Cell OFDMA Systems, June 2006

66. Z. Liang, Y.H. Chew, C.C. Ko, in *Proceedings of IEEE MILCOM'08*, Decentralized Bit, Subcarrier and Power Allocation with Interference Avoidance in Multicell OFDMA Systems Using Game Theoretic Approach, Nov 2008

67. A. Lotka, *Elements of Physical Biology* (Williams & Wilkins, Baltimore, 1925)

68. W.-S. Luk, T.-T. Wong, in *Proceedings of IEEE ICDCS*, Two New Quorum Based Algorithms for Distributed Mutual Exclusion, pp. 100–106, May 1997

69. A.B. MacKenzie, L.A. DaSilva, *Game Theory for Wireless Engineers* (Morgan & Claypool, San Rafael, 2006)

70. X. Mao, A. Maaref, K.H. Teo, in *Proceedings of IEEE GLOBECOM'08*, Adaptive Soft Frequency Reuse for Inter-Cell Interference Coordination in SC-FDMA Based 3GPP LTE Uplinks, Nov 2008

71. S.M. Mishra, S. Brink, R. Mahadevappa, R.W. Brodersen, in *Proceedings of IEEE DyS-PAN'07*, Cognitive Technology for Ultra-Wideband/WiMax Coexistence, Apr 2007

72. A. Mishra, V. Shrivastava, D. Agrawal, S. Banerjee, S. Ganguly, in *Proceedings of ACM Mobicom*, Distributed Channel Management in Uncoordinated Wireless Environments, pp. 170–181, Sept 2006

73. J. Mitola, Cognitive radio architecture evolution. Proc. IEEE **97**(4), 626–641 (2009)

74. J. Mo, H.-S.W. So, J. Walrand, Comparison of multichannel MAC protocols. IEEE Trans. Mob. Comput. **7**(1), 50–65 (2008)

75. R. Murty, R. Chandra, T. Moscibroda, P. Bahl, SenseLess: a database-driven white spaces network. IEEE Trans. Mob. Comp. **11**(2), 189–203 (2012)

76. J. Murray, *Mathematical Biology I: An Introduction* (Springer, Berlin, 2002)

77. J.W. Mwangoka, P. Marques, J. Rodriguez, in *Proceedings of IEEE DySPAN'11*, Exploiting TV White Spaces in Europe: The COGEU Approach, May 2011

78. H. Nan, T.-I. Hyon, S.-J. Yoo, in *Proceedings of IEEE DySpan*, Distributed Coordinated Spectrum Sharing MAC Protocol for Cognitive Radio, pp. 240–249, Apr 2007

79. M. Naor, A. Wool, The load, capacity, and availability of quorum systems. SIAM J. Comput. **27**(2), 214–225 (1998)

80. C. Perkins, E. Royer, in *Proceedings of the 2nd IEEE Workshop on Mobile Computing Systems and Applications*, Ad Hoc On-Demand Distance Vector Routing, pp. 90–100, Feb 1999

81. S. Pollin, M. Ergen, A. Dejonghe, L. van der Perre, F. Catthoor, I. Moerman, A. Bahai, in *Proceedings of IEEE CROWNCOM'06*, Distributed Cognitive Coexistence of 802.15.4 with 802.11, June 2006

82. F. Qian, Z. Wang, A. Gerber, Z.M. Mao, S. Sen, O. Spatscheck, in *Proceedings of IEEE ICNP*, TOP: Tail Optimization Protocol for Cellular Radio Resource Allocation, Oct 2010

83. M.I. Rahman, A. Behravant, H. Koorapaty, J. Sachs, K. Balachandran, in *Proceedings of IEEE DySPAN'11*, License-Exempt LTE Systems for Secondary Spectrum Usage: Scenarios and First Assessment, May 2011

84. S. Ramanathan, in *Proceedings of IEEE INFOCOM*, A Unified Framework and Algorithm for (T/F/C)DMA Channel Assignment in Wireless Networks, pp. 900–907, Apr 1997

85. T.S. Rappaport, *Wireless Communications: Principles and Practice* (Prentice Hall, Upper Saddle River, 2001)

86. B. Rengarajan, A.L. Stolyar, H. Viswanathan, in *Proceedings of CISS'10*, Self-Organizing Dynamic Fractional Frequency Reuse on the Uplink of OFDMA Systems, Mar 2010

87. S. Sengupta, S. Brahma, M. Chatterjee, S. Shankar, in *IEEE ICC*, Enhancements to Cognitive Radio Based IEEE 802.22 Air-Interface, pp 5155–5160, June 2007

88. S. Sengupta, R. Chandramouli, S. Brahma, M. Chatterjee, in *Proceedings of IEEE GLOBE-COM'08*, A Game Theoretic Framework for Distributed Self-Coexistence Among IEEE 802.22 Networks, Dec 2008

89. S. Shankar, C. Cordeiro, K. Challapali, in *Proceedings of IEEE DySPAN*, Spectrum Agile Radios: Utilization and Sensing Architectures, pp. 160–169, Nov 2005

90. S.J. Shellhammer, Estimation of Packet Error Rate Caused by Interference Using Analytic Techniques-A Coexistence Assurance Methodology. IEEE docs: P802.19-05/0028r1, Sept 2005

91. M.D. Silvius, F. Ge, A. Young, A.B. MacKenzie, C.W. Bostian, in *Proceedings of SPIE, Vol. 6980, Wireless Sensing and Processing III*, Smart Radio: Spectrum Access for First Responders, Apr 2008

92. J. So, N. Vaidya, in *Proceedings of ACM MobiHoc*, Multi-Channel MAC for Ad Hoc Networks: Handling Multi-Channel Hidden Terminals Using a Single Transceiver, pp. 222–233, May 2004

93. H.W. So, J. Walrand, J. Mo, in *Proceedings of IEEE WCNC*, McMAC: A Multi-Channel MAC Proposal for Ad Hoc Wireless Networks, pp. 334–339, Mar 2007

94. D.R. Stinson, *Combinatorial Designs: Constructions and Analysis* (Springer, New York, 2003)

95. A.L. Stolyar, H. Viswanathan, in *Proceedings of IEEE INFOCOM'08*, Self-Organizing Dynamic Fractional Frequency Reuse in OFDMA Systems, Apr 2008

96. A.L. Stolyar, H. Viswanathan, in *Proceedings of IEEE INFOCOM'09*, Self-Organizing Dynamic Fractional Frequency Reuse for Best-Effort Traffic Through Distributed Inter-Cell Coordination, Apr 2009

97. M. Strasser, C. Popper, S. Capkun, M. Cagalj, in *Proceedings of IEEE Symposium on Security and Privacy*, Jamming-Resistant Key Establishment Using Uncoordinated Frequency Hopping, pp. 64–78, May 2008

98. K. Sundaresan, S. Rangarajan, in *Proceedings of ACM MobiHoc*, Efficient Resource Management in OFDMA Femto Cells, pp 33–42, May 2009

99. V. Tawil, W. Caldwell, I. Reede, T. Kiernan, C.R. Stevenson, DRAFT 802.22a (Amendment) PAR. IEEE 802.22-09/0029r0, IEEE Std. PAR, Jan 2009

100. N. Theis, R. Thomas, L. DaSilva, Rendezvous for cognitive radios. IEEE Trans. Mob. Comput. **10**(2), 216–227 (2011)

101. M. Tokeshi, *Species Coexistence: Ecological and Evolutionary Perspectives* (Wiley-Blackwell, Chichester, 1998)

102. Y.-C. Tseng, C.-S. Tsu, T.-Y. Hsieh, in *Proceedings of IEEE INFOCOM*, Power-saving Protocols for IEEE 802.11-based Multi-hope Ad Hoc Networks, pp 200–209, June 2002

103. A. Tzamaloukas, J.J. Garcia-Luna-Aceves, in *Proceedings of IEEE ICC*, Channel-Hopping Multiple Access, pp. 415–419, June 2000

104. The Network Simulator (ns-2), http://www.isi.edu/nsnam/ns/

105. R. Vedantham, S. Kakumanu, S. Lakshmanan, R. Sivakumar, in *Proceedings of ACM Mobi-Com*, Component-based Channel Assignment in Single Radio, Multi-channel Ad Hoc Networks, pp. 378–389, Sept 2006

106. G.P. Villardi, Y.D. Alemseged, C. Sun, C.-S. Sum, T.H. Nguyen, T. Baykas, H. Harada, Enabling coexistence of multiple cognitive networks in TV white space. IEEE Wirel. Comm. **18**(4), 32–40 (2011)

107. V. Volterra, in *Variations and Fluctuations of the Number of Individuals in Animal Species Living Together. Animal Ecology*, ed. by R.N. Chapman (McGraw-Hill, New York, 1931)

108. F. Wamser, D. Mittelstädt, D. Staehle, in *Proceedings of IEEE VTC'10-Spring*, Soft Frequency Reuse in the Uplink of an OFDMA Network, June 2010

109. T. Wang, L. Song, Z. Han, Coalitional graph games for popular content distribution in cognitive radio VANETs. IEEE Trans. Veh. Technol. **62**(8), 4010–4019 (2013)

110. D.B. West, *Introduction to Graph Theory*, 2nd edn. (Prentice Hall, Upper Saddle River, 2001)

111. B. Wild, K. Ramchandran, in *Proceedings of IEEE DySPAN*, Detecting Primary Receivers for Cognitive Radio Applications, pp. 124–130, Nov 2005
112. WiMAX Forum, Mobile WiMAX—Part I: A Technical Overview and Performance Evaluation. Technical Report from WiMAX Forum, Aug 2006
113. S.-L. Wu, C.-Y. Lin, Y.-C. Tseng, J.-L. Sheu, in *Proceedings of International Symposium on Parallel Architectures, Algorithms and Networks, I-SPAN*, A New Multi-channel MAC Protocol with On-demand Channel Assignment for Multi-hop Mobile Ad Hoc Networks, Dec 2000
114. D. Wu, L. Zhou, Y. Cai, J. Rodrigues, in *Proceedings of GLOBECOM'12*, Energy-Efficient Resource Allocation for Uplink OFDMA Systems Using Correlated Equilibrium, Dec 2012
115. R. Yu, Y. Zhang, Y. Liu, S. Xie, L. Song, M. Guizani, Secondary users cooperation in cognitive radio networks: balancing sensing accuracy and efficiency. IEEE Wirel. Commun. Mag. **19**(2), 30–37 (2012)
116. Y. Yuan, P. Bahl, R. Chandra, T. Moscibroda, Y. Wu, in *Proceedings of ACM MobiHoc*, Allocating Dynamic Time-spectrum Blocks in Cognitive Radio Networks, pp. 130–139, Sept 2007
117. J. Yun, K. Shin, in *Proceedings of ACM MobiCom*, CTRL: A Self-organizing Femtocell Management Architecture for Co-channel Deployment, pp. 61–72, Sept 2010
118. X. Zhang, K.G. Shin, in *Proceedings of ACM MobiHoc'11*, Enabling Coexistence of Heterogeneous Wireless Systems: Case for ZigBee and WiFi, May 2011
119. H. Zhang, Q. Yang, F. Gao, K.S. Kwak, in *Proceedings of IEEE GLOBECOM'10*, Distributed Adaptive Subchannel and Power Allocation for Downlink OFDMA with Inter-Cell Interference Coordination, Dec 2010
120. J. Zhao, H. Zheng, G.-H. Yang, in *Proceedings of IEEE DySpan*, pp. 259–268, Nov 2005
121. H. Zheng, L. Cao, in *Proceedings of IEEE DySpan*, Device-centric Spectrum Management, pp. 56–65, Nov 2005
122. R. Zheng, J.C. Hou, L. Sha, in *Proceedings of ACM ACM MobiHoc*, Asynchronous Wakeup for Ad Hoc Networks, pp. 35–45, June 2003
123. R. Zhou, Y. Xiong, G. Xing, L. Sun, J. Ma, in *Proceedings of ACM Mobicom*, ZiFi: Wireless LAN Discovery via ZigBee Interference Signatures, pp. 49–60, Sept 2010

Index

© Springer International Publishing Switzerland 2014
K. Bian et al., *Cognitive Radio Networks*,
DOI: 10.1007/978-3-319-07329-3